Here Today, Gone Tomorrow

HERE TODAY, GONE TOMORROW

Environmental Travel Essays
for a Changing World

SHARON SNEDDON

ACKNOWLEDGMENTS

I would like to thank my editor, Lynn Post, and my book designer, Christina Dubois, for their excellent professional skills and patience in guiding me to the completion of this book. Thanks also to my associates in the local Sierra Club, Edmonds Climate Connection, and Save Our Marsh groups, along with Seattle Audubon for inspiring me with their tireless work toward our common goals.

HERE TODAY, GONE TOMORROW: Environmental Travel Essays for a Changing World

Copyright © 2018 Sharon Sneddon. All rights reserved.

No part of this publication may be reproduced, stored in a retrieval system, or transmitted in any form or by any means, electronic, mechanical, photocopying, recording, or otherwise, without the written permission of the author or publisher.

Library of Congress Control Number: 2018908462

ISBN 978-0-692-14103-8

First edition, 2018, printed in the United States of America

Published by Sharon Sneddon, www.heretodaygonetomorrow.me

Editor: Lynn Post, PostScripts Editing
Designer: Christina Dubois Publishing Services
Illustrator: Sharon Sneddon
Printer: IngramSpark

CONTENTS

INTRODUCTION

"A swallow-tailed gull has been seen the past few days on the breakwater at the Edmonds Marina. This nocturnal gull is native to the Pacific Coast off South America." The poster on the window of the tiny beach ranger station at the foot of the fishing pier featured a dark-headed gull with a charcoal back and white markings on its bill. I checked out the photos of this southern species, then headed out on the fishing pier, hoping it was still here.

The bulletin went on to say that this gull had been seen hanging around with the California gulls on the breakwater during the daytime since it feeds on squid at night. There were plenty of squid in our waters at that time.

Why was this South American gull way up here? Perhaps it followed some southern Pacific species of squid north as our waters warmed due to the second year of an El Niño. Perhaps this visitor was a scout for others that are making their way north as marine food sources head northward due to the warming ocean. Or perhaps this traveler just hitched a ride on one of the many freighters that ply our waters between the Pacific coast and the South American coast.

Anomalies like a southern species turning up far north of its traditional habitat are occurring more frequently. In April of 2018 a juvenile brown booby, usually found in tropical or subtropical waters off Central America, Mexico, and Baja California, was found, in weakened condition, on a beach in Newport, Oregon.

If marine and terrestrial habitats in the Pacific Northwest become too warm for some of our species and they move farther north, they will displace others already living in those more northerly habitats. Then where will those species go? This is just one of the questions scientists are investigating as changes in the natural world become more evident.

Our climate is on a wild ride right now, mainly due to the burning of fossil fuels that pollute our air, land, and water and contribute to global warming. We need to take a hard look at our lifestyles, expectations, and traditions and make some drastic changes before it's too late.

With world population approaching nine billion, we need to determine how many people this earth can actually support and examine how the natural resources all of us depend on are being exploited. Increases in revenue for companies destroying our natural resources should be held in check before the loss of those resources upsets the balance of our ecological systems.

Instances where lifestyle or cultural traditions are contributing to the demise of natural resources should be adapted to conditions as they are today, not in the past, when those traditions began. The world of our ancestors no longer exists.

Is global warming already having an impact on our planet and, if so, how do we know? How did those who came before us survive droughts? Are there life skills we have lost that guide animal societies to survive today? These topics are just a few of the many highlighted in my essays written from personal experiences in many parts of the globe and in my own backyard.

Why did Costa Rica's golden toad disappear, with other amphibians worldwide following suit? Will rising sea levels destroy the ancient totem poles on Haida Gwaii off the west coast of Canada? The Quileute Tribe on the west coast of Washington state is already making plans to move upslope into the Olympic Mountains before sea-level rise floods their small coastal town.

What about the spread of insect pests such as the fire ants that nipped at my toes in the British Virgin Islands? The thought that they are marching northward, toward my home state of Washington, is frightening.

Over twenty-five years ago I joined Seattle Audubon and began going on some of their field trips. Those excursions opened up a whole new world to me. I was already nurturing my new-found interest in nature photography; birding added an extra dimension. Trips to other states or countries always had an element of birding.

I wasn't interested in just taking close-ups of birds, but wanted to get an image of the bird in its habitat. The more I learned about habitats, the more I discovered that so many of them were changing or disappearing. Many of these birds were under stress from our expanding developments, explorations for oil and gas, pollution, and other human-caused habitat disturbances.

My hope is that this collection of experiences will compel you to take time to form a connection to the natural world, then take steps to preserve and protect what we have. Observe, listen, watch what is happening right now, in this place. This is a fleeting moment; we can learn things here. Turn off the cell phone, stop texting. Technological communication devices will only become more advanced to pull our minds deeper into their world and farther away from where we are physically at this moment. Once you shed the predictability of algorithmic communication systems, social media, mobile phones, tablets, etc. that allow us to be connected to others every instant of our lives, you open up space to experience the present moment. Pay heed to those moments, listen to the echoes of the earth urging us to protect our home and all the creatures living here.

The prose poems in the first section, "Brief Encounters," flowed onto my notepad without much thought; just becoming aware of my immediate environment. The essays in the remaining sections were filtered through my mind carefully to convey my assessment of a situation I was in at that moment and determine what I could learn from it.*

---

*Variations of a few of these essays have been previously published elsewhere.

Daily news of climate change gets overwhelming. People fall into a state of inertia, not knowing what to do that might help slow down some of these changes. Yet, many aren't aware that they can influence larger events to create a better world. Sometimes it just takes a shift in attitude, a bit of introspection. What is important here? Efforts all of us make, no matter how small, can make a big difference in the long run. Take small steps, but start taking them now.

Write a letter to the editor of your local newspaper to raise awareness of an issue for other readers (instructions for writing a letter to the editor are on my website: www.heretodaygonetomorrow.me). Urge your legislators to support environmental issues you care about. Join a local environmental group. Look for actions you can take close to home. We can all make small changes; we can all speak up to protect the earth—our home.

BRIEF ENCOUNTERS

A FEATHERY EMBRACE

The beach is deserted this blustery April afternoon as I stroll along the sand, just out of reach of spreading tendrils of foam washing ashore. All of a sudden, I feel a breeze at my back, now dividing into two feathery columns rushing within arm's length on either side of me, then reforming into a pulsating cloud of sandpipers dancing on the wind as they follow the water's edge.

Halting my beach walk, I watch as the flock unites into a V formation once again, then disappears over the pounding surf. This close encounter with a fragile, long-distance migrant gives me a chill. Here briefly to rest and refuel on a fourteen-thousand-mile journey from their winter home in Argentina to Arctic breeding grounds, they still have a long way to go. So close, I could have touched them.

About thirty feet ahead of me, three shorebirds probe the mud with their long bills. Getting closer, I see two black-bellied plovers, elegant in their black-hooded capes, feeding alongside a busy dunlin, scurrying across the sand. The three hurry this way and that, probing, hustling to a new spot, probing again.

Out of the corner of my eye, I spot a cat-size, furry, gray animal bounding through the windblown beach grass. The rabbit stops, turns his long ears to listen for predators, then disappears.

Cascades of white foam are thrown toward the beach by violent breakers that tear apart the churning sea, glistening in the late-afternoon sun. Five brown pelicans skim along just above the surf, like miniature airliners flying in close formation.

Then, in an instant, all are gone. I'm alone on the packed, wet sand. Just me and the sea.

Cattails Waiting

Withered and bent from winter's wet chill, the cattails around Green Lake wait solemnly for spring. Amidst a burst of breeding activity, red-winged blackbirds will weave nests of grasses onto the bottoms of cattail stalks, where they are sturdy. There they will be pushed to and fro as a male red-wing chases away his competitors, then flies off to mate with another female as the whim arises.

Soon tiny male marsh wrens will enter the lakeside scene, each building several nests out of grasses, cattails, and sedges, giving the female a chance to choose one. The chosen nest will be formed into an oval shape with an opening on the side and anchored to the cattails a few feet off the ground.

Just beyond the cattails, rafts of coots that glide along the edges of the lake year-round, will begin constructing a floating platform nest from the stems of marsh plants, well hidden in vegetation.

Ducks aren't pairing up yet, but will be soon. Now, a female common merganser swims with a coot while a goldeneye and a male merganser dive together, searching for lunch.

All the resident mallards and coots have shared their watery home this winter with visiting waterfowl: goldeneyes, mergansers, and widgeons with their rubber-ducky sound. Soon those visitors will head for their own nesting grounds on the eastern side of Washington's Cascade Mountain Range.

Around the northwest side of the lake, the steel-blue waters undulate in a corduroy pattern. Most of the faded-denim-colored water is still on this chilly, sunny winter day. Mud and early alder buds scent the air.

By late spring and early summer, the cattails will share the lakeshore with yellow pond irises. A gentle breeze will nudge along small sailboats, kayaks, and pedal boats as a parade of human and canine walkers briskly pass the cattails, unaware of the busy avian activity hidden within the slender stalks.

CELEBRITY SEAL

"Look, Dad, a seal!" shouts a young boy, all bundled up on this cold January day. The boy and his father hustle down the boardwalk to the fishing pier, where they join five others peering over a wooden railing. The attraction is a juvenile harbor seal hauled out on a concrete walkway, head jutting out above the calm waters. The seal's black button nose points briefly toward boats moored ahead of him, then he flips on his side. Cameras click at his every move.

Clad in form-fitting gray-and-tawny short-haired fur, the seal appears quite chic. Unconcerned by all the camera lenses pointing his way, he lounges on his stomach with flippers snug on each side of his plump body. Occasionally he moves the ends of his flippers from side to side in a teasing manner, then glances into the mirror-smooth water as if admiring his dreamy eyes and long, handsome whiskers.

This ambassador for his species has made a daily appearance on the waterfront walkway of our town for a month. His photo on the front page of our local *Beacon* newspaper gave a boost to his celebrity status. He seems to enjoy being in the limelight.

I hope he sticks around awhile. A disconnect from nature plagues our society in this twenty-first century. Even watching a lone seal on a concrete walkway provides a thread to the natural world.

MYSTERIOUS MURRELETS

What was that crackle? Perhaps paw steps in the ferns about twenty feet from the trail I'm on? Black bear and mountain lion are here. Maybe.... The dirt road leading to Stout Grove here in Redwood National Park is so narrow, with huge redwood trees on either side, there is barely enough room for my car to pass through. Not much room for cars, but plenty of secret spaces for furred or feathered creatures.

Later I amble along a trail in Lady Bird Johnson Grove, passing ancient conifers soaring into the pale blue sky. No fog or mist dampening the dusty trail today. Branches droop, as if trying to reach into the earth for the moisture it holds. Late summer heat has baked some of the ferns, leaving fronds brittle and broken. Bits of bright-green moss and greenish-white algae litter the dirt path.

The sign at the trailhead features an illustration of a marbled murrelet. Visitors are asked not to litter bits of food in this area. Jays, crows, and ravens are attracted by the food and may become aware of murrelet eggs or hatchlings nearby, an easy snack. The apple I was nibbling gets stuffed back into my pocket.

A small, chunky bird, the marbled murrelet spends most of its life on the sea, yet flies up to fifty miles inland looking for the perfect branch in the perfect old-growth tree upon which to lay its eggs. I know I won't see one but imagine I can feel their presence. That's enough for me.

Owl Eyes

BURROWING OWL

Yellow eyes gleaming, a burrowing owl watches the wheels of our van pass about fifteen feet in front of it. Close-cropped tawny feathers enrobe the striped owl, who stands guard near its burrow at the edge of a farm field. Borrowed from a ground squirrel, the underground home may shelter owlets come spring.

Danger lurks everywhere for the owl, even under the noonday sun. Not far down the road, as we stop to scan a field for western meadowlarks, three abandoned dogs—two black and one rust-colored, with ribs showing—approach the van, tails wagging, desperately hoping for a handout. Closer to the burrow, a pile of coyote scat hardens in the dry air.

GREAT HORNED OWL

So close we don't see it above us, a great horned owl patiently watches yet another parade of hikers filing along the path through its forest home. Perched on a hefty sycamore limb in full view, the owl closes one eye, looking bored with the whole scene.

If it were spring and the owl was guarding a nest of fluffy owlets, watch out! Those sharp talons can grasp a rat, a rabbit, or other intruder in a flash.

The eyes of both of these owls seemed to say to me, "This is my home; you do not belong here." They're right. Let's leave some places for the owls; it's the least we can do.

Rituals

Thorny vines grip my bare arm as I pluck off a plump, juicy blackberry, deep purple as a fading summer sunset. I drop it in my white plastic bucket on top of the growing pile of these sweet, sensuous gifts of this sunny season.

The scraggly patch of vines edging Scriber Creek, about a half mile from my home, does not grow peacefully in a secluded wood. Thinly stretched between a busy city street and a dismal strip mall, the tenacious vines are ignored by the city mower each fall.

Crumpled plastic bags, along with discarded paper Big Gulp cups, litter the ground in front of the vines. Here and there are faint hints of animal trails through the weeds leading up from the creek behind the vines and ending on the short brown grass where I stand. Muskrat or opossum, most likely.

Overhead behind me, near the wooden bridge over the creek, a juvenile barn swallow nags its streamlined blue-and-rufous parent for another meal. A couple of months ago, those macho, square-shouldered aviators dive-bombed me when I strolled over that bridge, where they had carefully constructed their mud nests under the boards.

A yellowjacket streaks in and circles my bucket until I give in and set it on the ground. Satisfied after sampling a few, the yellowjacket works its way down to a sunny patch of vines ahead of me. We are both greedy, this yellowjacket and I, both drawn to the most succulent bunches.

Picking blackberries at the end of summer is a ritual for me, an anchor to the steady rhythm of nature. I find that simple seasonal rituals—like raking leaves as I listen to squads of Canada geese winging their way south; walking in fluffy new snow on a velvet winter evening; or picking blackberries at the end of summer—have become touchstones on the turbulent sea of life. These transitions of the seasons are always there; I can count on them, if nothing else.

I poke my nose into the nearly filled bucket, engulfed in the sweet scent. That's enough for me; I'll leave some for the other creatures who rely on them more than I do.

Shoveler's Pond

The trill of a red-winged blackbird beckons me to Shoveler's Pond. Now filled with winter rain, this vernal pool will dry up by midsummer. My boots sink into the soggy soil at the pond's edge, where a few small willows have found a foothold, bright-green catkins hanging from their spindly branches.

In the center of the pond, a scraggly tree about thirty feet tall has lost its grip in the earth, flopping down horizontally onto a mat of dried reeds, as if taking a winter nap. A few feet from the western edge of the pond, three female northern shovelers tip headfirst into the still water in search of a meal; opportunists taking advantage of a temporary habitat on their long migration north to their summer breeding grounds.

Snow Geese

First the whir of thousands of wings as they rise in the air, responding to some ethereal call. Cries erupt from the loose flock as they circle overhead, then alight in a field farther away from McLean Road, here in Washington's Skagit Valley.

What stirred them? A photographer getting too close? A pickup heading down the road? A peregrine falcon overhead? An alarm call, wing swoops, momentum tying the group together, following each other until a cloud of white blurs streams overhead amidst a deafening roar, heading toward an even farther field.

Approximately twenty thousand snow geese overwinter in the valley. That's partially due to farmers who are paid to plant grain on a portion of their land for these winter visitors. But it's not solely a benevolent gesture. The larger the snow goose population, the more geese for hunters to gun down October through mid-January.

Hunting season is over now, so the three thousand to four thousand in these farm fields, along with thousands more in other fields, can feed and rest in relative peace. There's a long journey ahead to their breeding grounds on Wrangel Island in the Russian Arctic.

What a joy to welcome them each fall, like my family coming home. Their calls tug at my soul, pulling it up into the air. Take me with you!

WABI SABI IN THE DESERT

Graceful, aged beauty lying weathered on the desert soil. *Wabi sabi,* the Japanese term for "old, graceful, lovely," seems to fit perfectly. Twisted limbs of the old saltbush appear exhausted from the heat and wind that bakes a land where only the toughest survive.

Round-tailed ground-squirrel burrows appear here and there, providing a cool home underneath the parched soil. A pile of dark-brown scat, probably left by a coyote, hardens at the base of an ocotillo.

Scraggly mesquite offer a perch for winged desert dwellers. Black-chinned sparrow and greenish verdin flit among the brittle bushes, catching insects in the cold, dry December air. A Cooper's hawk keeps watch from a light pole by the edge of the road.

Spent red and purple buckshot shells and empty beer bottles testify to human presence—a late-night party in the desert. Which furry or winged creatures were targets for that lead shot?

Revelers long gone, all is quiet now out of respect for the deceased saltbush, resting at last on the desert floor.

ON HOME GROUND

AVIAN INVADERS

A flash of rusty orange, green, and white darts from my hanging fuchsia basket as I open the sliding glass door to the deck. The rufous hummingbird is gone in an instant. Yay! They're back!

Less than four inches long with a wingspan just over four inches, these winged jewels brighten summer gardens for many people living in the Pacific Northwest. But, where do they go in the winter? I can't imagine this tiny guy migrating any farther than my neighbor's yard.

Weighing in at less than a quarter of an ounce, this feisty fellow makes a two-thousand-mile round-trip migration each year. In early spring, they usually leave their wintering grounds in Mexico and southern Texas and fly to their breeding grounds along the Northwest coast, some traveling as far as southwestern Alaska. In recent years, some have been wintering in the Gulf States. The males arrive in the Puget Sound region in late February through mid-March. Females arrive a week or two later. They'll remain here until late summer and early autumn, when they head back south.

Returning to our area in the spring, rufous is entering the territory of our resident Anna's hummingbird, where they must compete for food. These bossy birds sometimes engage in serious wing-to-wing combat at feeders!

Anna's hummingbirds lived year-round in northern Mexico and southern California until about forty years ago, when they expanded their range north to include Washington. "Apparently [Anna's] has a more favorable ability to withstand colder temperatures and is better adapted to the introduced winter flowering plants we have here," explains Brian Bell, Master Birder at Seattle Audubon.

More and more feeders filled year-round are an additional enticement. "Except in the most extreme circumstances, hummers would probably survive just fine without feeders and very likely get only a fraction of their daily food from feeders," explains Dan Harville, who taught a class on hummingbirds.

Though hummingbirds are attracted to red, since many of the flowers where they find nectar are red, don't add red food coloring to the sugar water because it may be harmful to them. "Feeders are almost always red themselves," says Harville. "If the feeder is somewhat under cover or out of the way, attaching a red streamer to the bottom of the feeder should help."

Each year more and more birds, plants, and animals are making their way north as a warming climate changes their home turf. Changes in the plant communities due to the warming climate seem to be especially affecting wintering habitats. According to Marianne Taylor in her 2016 book *Nature's Great Migrations*, some rufous hummingbirds that typically winter in western Mexico are wintering instead in the far southeastern United States due to changes in their food sources.

Unfortunately, some bird and animal species here in the Northwest are also moving north as their habitat warms, food becomes available earlier or later than they are accustomed to, and newcomers are competing for limited resources. As people level forests, fill wetlands, and pave and develop every available acre, birds and other animals will have nowhere to go. At that point, we will begin to lose some species.

Losing species will begin to unravel the web of life that we humans depend upon. It's too late to convince some of these early invaders to stay where they are because that habitat is changing too. We must deal with a warming planet now, for the survival of all species, including our own.

You can help by planting native species in your yard, providing a water source, and leaving some brushy areas so birds can find cover. Tidy yards are out!

COPING WITH CHANGES

"I would hate to have all our customs, this history, lost," said Makah elder Helen Peterson when I spoke with her many years ago on the reservation in the far northwest corner of Washington state. One of twelve elders of the tribe at that time, she had devoted her entire life to teaching young Makah tribal customs, including basketry, storytelling, and dances. Many scholars viewed her as a storehouse of tribal knowledge. Peterson often represented her tribe at Native American cultural events, including one held at the Smithsonian.

On the day I spoke with her, she was in a classroom full of third graders eager to learn about their culture. The few Caucasian students from the local Air Force and Coast Guard bases listened intently as well.

Along with traditional songs and dances, she tells ancestral legends to instill some Makah values into these eager learners. The morals embedded in their legends teach young people how to live a traditional life. One of her favorite legends was about Thunderbird:

Thunderbird is a giant Indian living on the highest mountain. Whales are his food. When he is hungry, he puts on the head of a giant bird, a pair of giant wings, and ties Lightning Fish around his waist. Lightning Fish has a head as sharp as a knife and a red tongue that spews fire. Thunderbird's huge wings darken the sky, making loud noises as they move. Seeing a whale, he throws Lightning Fish into its body to kill the whale. He then carries the whale back to the mountains to eat.

Salty breezes on this windswept reservation remind me of the sea nearby, with legends often stressing the Makah connection to the marine environment. Throughout the past, fishing, sealing, and whaling were mainstays of their economy. Fishing is still a mainstay in their diet, but there is no longer a subsistence need for sealing or whaling. Traditional foods included seal, whale, salmon, halibut, snapper, rock cod, camas root, salmonberry sprouts for green, and shellfish such as clams and oysters. "Now we eat beef, chicken, turkey, salads, cakes, modern stuff," Peterson commented.

In 1855 the Makah ceded thousands of acres to the Washington territorial government. The treaty between the United States and the Makah Indian Tribe preserved the Makah's right to take fish, whales, and seals at their usual and accustomed grounds. Fish were plentiful in the ocean and rivers in those days, but abundance of that food source is dwindling due to warming waters, invasive species, and new pests. Indigenous people, like the Makah, are speaking out about the dramatic impacts to their way of life.

In an article in *The Seattle Times*, April 10, 2018, Timothy Greene Sr., former chairman of the Makah Tribal Council, reminds the scientific community that the inherent knowledge of the land and sea held by indigenous people needs to be considered along with science when searching for ways to address climate change.

With temperature changes in the waters of the Pacific resulting in fewer of the fish they depend upon and other significant changes to natural areas where the Makah hunt and gather plant foods, the tribe is calling on decision makers, who are not experiencing these impacts in their backyards, to make changes now. Rising sea levels and more frequent fires and flooding are already affecting the tribe. A decrease in an economically valuable fish such as whiting (hake), used to make fish sticks, impacts the Makah.

"For example, stocks of many fish species, like Pacific hake, are sensitive to ocean temperature along the California Current, and recent declines in their numbers have serious implications for the well-being of my own Makah Tribe," states Greene.

According to a report issued in 2016 by the Northwest Treaty Tribes, climate change is affecting treaty rights and their own way of life. Disruptions to usual behavior include marine fish moving away from historical fishing grounds to cooler water north and invasive species from the south moving north to compete with native species already living in those habitats. That situation has also affected wild plant foods and wild game, which are moving north or to higher ground due to changes in their environment.

The Makah, along with other Northwest tribes, are taking the long-term view when planning for natural-resource management, considering the future implications of drought, invasives, and wildfires. Tribal members and their partners are protecting intact ecosystems and repairing damaged ones. Education within the community, especially the youth, regarding resilience to these changes, plays a key role in their adaptation.

All too often the knowledge of indigenous people has been disregarded by the scientific community. We need all voices now, speaking out loud and clear to tell the fossil fuel industry that we must make the switch to clean energy before more damage is done.

"While others debate the causes of climate change," maintains Greene, "we who live close to the land are experiencing the major impacts and call for immediate and strong action to protect the resources on which we all rely."

CRAFTY CORVIDS

Gliding over the treetops, a lone crow steadies its wings so as not to break the silence of the morning. Where I walk on a bluff above Puget Sound, the air is already getting muggy this August day.

Below me, two crows search the gravel next to the railroad tracks along the shoreline. All of a sudden, one crow rises straight up in the air and lets something drop onto the gravel. Then it swoops down to scoop up a snail from its crushed shell. Slime drips out of the crow's bill while the smaller crow watches intently. Teacher and student; this young crow is learning a survival skill. Adult crows pass on to juveniles knowledge gained through experience, reinforcing the strong bonding in their social groups.

I have read that they are very gregarious, living in family groups of five to nine birds. On walks around my neighborhood, I often see several of them feeding in the same area or just squawking at each other. A mated pair stays together for life, with older offspring sometimes acting as helpers in raising their younger siblings. Maybe those noisy birds were having a little family tussle.

Nearly anything makes a meal, with food inside wrappers or plastic bags fair game. These astute avians store snacks in a variety of places and don't forget where they are stashed—under the bark of a tree, under a bush, under a shingle on a roof.

A tasty treat might include bugs, worms, insects, mice, berries, carrion, roadkill, burgers, fries, or other finds. They also prey on the nests of songbirds, eating the eggs and young. I wonder if that's who dispatched a beautiful juvenile pine siskin I found on the driveway. Picking it up between two pieces of cardboard from the recycle bin, I placed it under a bush, noticing the deep gash that had smashed the back of its neck.

Even more disgusting are accounts from the Middle Ages (400–1500 CE) that describe how flocks of crows would often gather before a battle, knowing they would get a free meal from all the dead bodies after the combat.

Skilled at calculating the speed of an oncoming vehicle, crows dining on roadkill sense just how long they can continue indulging before they become roadkill themselves. In Port Townsend, Washington, locals report watching crows drop chestnuts in a busy intersection, knowing cars will crush the shells. In the southern United States they have been seen hanging upside down, shaking branches to release spiders, insects, and small lizards to other crows waiting on the ground below.

Crows are found all over the world except in New Zealand, Antarctica, and South America. A researcher from Massey University in New Zealand reported observing crows in New Caledonia making two different kinds of stick tools to forage for invertebrates such as insects, centipedes, and larvae. After they have made their stick tool, they will carry it around to use at different locations.

Why search for food if you can swipe it from someone else? To steal a fish from an otter, one tactic is for one crow to distract the otter by pulling its tail while another grabs the fish from its mouth. A crow intently watched one morning as a squirrel buried a nut under a tree in the neighbor's yard. As soon as the squirrel left, the craft corvid swooped down and claimed his prize.

Crows are playful. Antics include dropping a stick from on high to see if they can catch it before it hits the ground or sliding down a snowbank on their backs, climbing back up, and sliding down again. They have been known to sneak up and tweek the tail of a sleeping dog, then quickly fly off.

These sleek, swaggering scavengers possess a varied vocal repertoire including alarm calls, distress calls, scolding calls, and assembly calls.

Come autumn, in the late afternoon, I often hear large flocks overhead, calling to each other as they head to their roosting place for the night.

"Safety in numbers," muses John Withey, researcher at the University of Washington, in Seattle. "Local roosts are often in or near water or wetlands, perhaps to help avoid terrestrial predators, like raccoons," he continues. "They also provide possible social learning opportunities, such as where to find potential mates. They all leave in the morning, so the actual roost area doesn't seem to be used for feeding."

It would seem to me that watching wildlife, even crows, can provide insights into what our culture may be lacking that holds society together for other species. Learn from each other, work together toward a common goal. Good words to live by.

EDMONDS MARSH

The cheery notes reach my ears before I spot the marsh wren gripping a cattail, head thrown back in joyous song. A long tail provides support on the reed as the wren, like a diva on stage, announces the coming of a balmy spring day to the audience.

Beyond the songster, cattails march across the freshwater zone toward small muddy ponds in the distance. Red-winged blackbirds provide the cattail chorus while, at the edge of a pond, five slinky great blue herons act out their shady-character roles.

One of only a few saltwater marshes along Puget Sound, the 22.5-acre Edmonds Marsh hosts an array of wildlife along with resident birds and others migrating through. Neither resident nor migrant birds seem to mind the rumble from trains about thirty feet from the western edge of the marsh. They seem oblivious to the barking in the play yard at the Blue Dog Doggy Day Care Center. The heady aroma wafting from Gallaghers' brew pub doesn't seem to faze them either.

The scene is different each time I visit. Spring paints the grasses a vibrant green; there is a freshness in the air. Tall magenta spikes of purple loosestrife (an invasive plant) decorate emerald grassy patches in July, while pale gold tints the autumn when cattails wither, bending down to lie on the mud, weary from long days in the sun. A stark brown beauty reigns through the winter, when the bare branches of the alder reveal their true form, as a chilly wind rustles feathers on skulking blue herons.

Less than one half the size it was originally, it still provides unique services for our human environment. In addition to providing habitat for wildlife, the array of benefits provided by the marsh include filtering pollutants from the surrounding land as well as stormwater and flood-water retention.

Here and there along the edges of some of the ponds, slick rainbow colors rim the mud, a legacy of its slippery, smelly past when Unocal used the site for a bulk fuel-oil terminal in 1926 through 1991. The site was deeded to Edmonds, a city of forty thousand, about fifteen miles north of Seattle.

Union Oil Company of California (Unocal) continues working on remediation of the land and water, but other hazards remain. The threat of toxic coal dust from open railroad cars as well as explosive crude-oil tank cars travel the rails just to the west of the marsh. Up to forty trains per day travel north along this route moving passengers and commodities, including coal and oil. That number is likely to jump to one hundred trains per day by 2030. Coal dust blowing off the railcars lands in this fragile landscape daily. An oil spill would be devastating to the marsh and the nearby shoreline; an explosion would be catastrophic to the whole region.

Two creeks, Willow and Shellabarger, flow into the marsh, then through drainpipes into Puget Sound. Prior to the drainpipes being installed, coho and Chinook salmon from nearby Puget Sound took advantage of the quiet waters of the marsh for resting and feeding. Now there are plans to replace the drainpipes with open culverts that would be more inviting to daylight-loving salmon and less likely to collect sediment, which causes flooding of nearby streets during heavy rainfalls.

Who lives in the marsh? Nearly two hundred species of birds use this unique ecosystem, some year-round, and others for resting and refueling along their migration routes. I've seen swallows and warblers in the spring zipping through the alders. Green-winged teal, Northern shoveler, and gadwall often take refuge there during winter months. Mallards usually show up year-round.

The mudflats provide a smorgasbord for migrating shorebirds, such as long-billed dowitcher, semipalmated plover, and dunlin. Killdeer are permanent residents, skittering along the mud in search of food. There are other stealthy, slithering, slinking critters, such as long-tailed weasel, muskrat, and raccoon, all adept at hiding in the reeds. Western salamander, garter snake, and red-legged frog also take advantage of this swampy landscape.

What is our responsibility for this fragile wetland? Our moral imperative, it seems to me, is to protect, preserve, and attempt to restore this unique habitat as close as possible to its original form. Maintaining a scientifically determined buffer zone around the marsh, protecting it from encroaching development, is extremely important.

According to a U.S. Geological Survey (USGS) report, established methodologies to reconstruct wetlands successfully do not currently exist. Wetlands contain extremely complex systems that provide many services to many species. In other words, once it's gone, it's gone.

Some specific restoration efforts are paying off. Volunteers with EarthCorps and Friends of the Marsh are removing invasive plants, such as Himalayan blackberry and purple loosestrife, and replanting native species. According to Keeley O'Connell, restoration ecologist, "Edmonds Marsh is a unique asset to the city as it provides a space for local and migrant birds to nest and feed and filters rainwater running off nearby streets before it drains into Puget Sound."

These measures, along with the new salmon-friendly open culverts, are a step in the right direction. The City of Edmonds has recently hired an environmental consulting firm to produce a plan for restoring and improving the marsh. But the singing marsh wren cares not what was before or what surrounds this habitat now. The sun is shining—it's a glorious day.

GAINING STRENGTH AS SHE SLUMBERS

Sitting serenely on a bed of puffy clouds, Mount St. Helens slumbers outside my Alaska Airlines window. Dark shadows play among the ridges and crevices around the rim of the crater. A light blanket of snow invites this quiet slumber, getting some rest before the next big show.

May 18, 1980, at 8:32 A.M. was the last grand performance. I was clinging to handrails that breezy morning, on my first sailboat outing on Puget Sound. "What was that?" someone shouted as a thunderous boom echoed around us. "A sonic boom?"

"Maybe it was Mount St. Helens," came another reply. "The mountain's been having small tremors for weeks now."

Mount St. Helens it was. An earthquake registering 5.1 on the Richter scale had awakened the sleeping giant, unleashing ash clouds along with huge lahars (mudflows) that were destroying everything in their path for 150 square miles on the northwest side of the mountain. Heat from the blast reached 680 degrees, sending a searing mass of ash and mud that wiped out fifty-seven people. Though scientists had been predicting an eruption at any time, they underestimated its size.

Old-growth forests standing for centuries were flattened. Thousands of deer and elk were smothered in the hot ash along with two hundred black bears as well as millions of birds and small animals. Hearing details about the eruption on the evening news that night, I wondered how anything at all could have survived such a blast.

Days after the eruption, state, county, and local officials were coping with an ever-expanding danger zone, while homeowners at Spirit Lake, on the southeast side of the mountain, were hounding them for access to their properties, wondering if anything remained.

According to Steve Olson in his book, *Eruption: The Untold Story of Mount St. Helens*, Weyerhaeuser, the timber company, played a key role in the high death toll on that day. For nearly one hundred years, the company had a history of continuing with scheduled logging, even in the presence of danger. Mountain about to erupt or not, they were set on finishing logging some old-growth acreage a few miles from the rumbling, shaking epicenter of the mountain. Danger in the air, many of the loggers wanted to stop the operation and leave the area. Some did. In the end, the danger-zone boundaries were set by authorities according to Weyerhaeuser's logging plan for the day, which was way too close to the epicenter. Many of the deaths could have been prevented, if Weyerhaeuser hadn't been set on logging that area.

Itching to drive the 150 miles south from my home in Seattle to see firsthand the results of this epic event, I finally got the opportunity several years later. Highway 514 winds east from the town of Castle Rock around the north side of the mountain. Huge mountainside scars of downed trees were shocking, but seeing miles of hardened mudflows was ominous. I couldn't imagine the terror the people and the animals felt watching that wall of hot mud and ash bearing down upon them.

Ash, lava, and debris clogged the Toutle River west clear to Interstate 5. Towns in the eastern part of Washington state were plunged into darkness as ash filled the air, canceling air traffic over the state. Winds carried the ash across the country. One positive outcome came as a surprise: the ash made good fertilizer. Farmers in eastern Washington were pleased, while folks on the western side of the Cascade Mountain Range, and, especially near Mount St. Helens, had to deal with the destruction and cleanup.

Luckily, it was early spring when the mountain erupted, so some of the mice, voles, shrews, pocket gophers, and other small creatures were still in their underground burrows. After the ground cooled, they set about digging for roots and bulbs to eat, mixing the ash (tiny particles of rock and glass) with dirt, scattering seeds in their droppings. The wind carried in more seeds, insects, and spiders.

Prairie lupine, fireweed, and pearly everlasting were some of the first plants to take root, followed by thistle and sedges. When these plants died, they provided food for other organisms to recolonize the area. Within a few years, conifers began to reappear. It seemed part of the natural cycle; pockets of life left after the eruption fed into the renewal of the devastated area.

Wanting a more personal experience, I phoned my recently widowed Uncle Boyd living in Vancouver, Washington, about eight years after the eruption. He and Aunt Betty had a cabin at Spirit Lake for many years. "I'll meet you in Woodland, and we'll take a drive around the south side of the mountain to Spirit Lake," he eagerly suggested.

That morning, we drove northeast on Highway 503 from Woodland, through the tiny town of Cougar, stopping to survey acres of "ghost forest," where broken, bleached trees were all that was left of the lush forests. Farther up the road, we came to a small community of homes, mostly A-frames, nestled close to Spirit Lake. Fortunately, the lava and mudflows hadn't reached many of the cabins, including my uncle's. Melancholy tinged his recollections of cross-country skiing in the winter along with joyous family get-togethers on weekends.

In photos taken just after the eruption, Spirit Lake looked as if someone had emptied a giant box of toothpicks onto the murky brown water. Now it shimmered a deep azure. Aunt Betty's ashes were scattered on the waters of Spirit Lake a few years before my visit and, when he passed away several years ago, Uncle Boyd's ashes joined hers in that peaceful setting.

From total devastation, Mount St. Helens is transformed into one of the most biodiverse places on Earth today—because nature was in charge. In today's Anthropocene era, where humanity is influencing climate change, both land and waterscapes are experiencing rapid changes which will likely have catastrophic results much different than the eruption.

According to the U.S. Geological Survey, volcanic eruptions can impact climate change, but not to the degree humanity can. During the eruption of Mount St. Helens, carbon dioxide, a greenhouse gas, was released, but it was less than 1 percent of that released by human activities. In 2010 alone, there was an increase of thirty-five billion metric tons, most of that from burning fossil fuels. While the mountain released ten million tons of carbon dioxide in nine hours, human-caused release is constant and steadily increasing.

How will the area appear to visitors in 2050 and beyond? I pondered this question as I loaded a backpack into my car one August for a trip to view the summer wildflowers on the northwest side of the mountain. Focusing my Nikon on the glorious colors decorating the dry, brown grass, the fury still within the volcano was the last thing on my mind. Still, I couldn't help but wonder whether I'd make it out alive if the mountain decided to wake up that day....

JUST PASSING THROUGH

So colorful, such graceful flyers,
the kites;
So colorful, skittering along the shore, just ahead of the foamy surf,
hordes of kids;
So colorful, scattered over every inch of the beach,
sun umbrellas, beach chairs, picnic coolers.

Where are the brown pelicans, sanderlings, pelagic cormorants? So many people have descended on the beach this unusually hot, humid day, I give up scanning the waters near the jetty for the spray from the blow of one of the resident gray whales. People on the shore, people on the jetty, people in the water.

The wind, always the wind, providing an aerial highway for these avian migrants on their way south in the fall to their wintering havens, then north in the spring to their breeding grounds. This six-mile-long, two-mile-wide peninsula jutting into the Pacific from the lower Washington coast is on the Pacific Flyway for migrating birds such as western sandpipers, black-bellied plovers, ruddy turnstones, and many others each spring and fall. Ocean Shores also attracts lots of tourists.

Wide sandy beaches border the peninsula on the west side. A rock jetty at the southwest corner extends about two hundred yards into the sea, connecting with a rock retaining wall extending just over a mile east of the jetty. With waves lapping at the rocks, the jetty is one of the best places to spot birds like black oystercatchers, with their long legs and striking red bills. None here today.

Ocean Shores isn't normally this crowded; that's why I can usually count on a relaxing, refreshing getaway, just a three-hour drive from my home near Seattle. But today, I'd guess at least around five hundred of the three million annual visitors are here this warm weekend.

People everywhere. It's hard to believe that any wildlife at all could survive around here. But according to staff at the Ocean Shores Interpretive Center, at least six black bears, several cougars, bobcats, black-tailed deer, and other creatures roam the patches of forest in the interior of the peninsula.

With unpredictable weather patterns becoming the norm, many birds are arriving a couple of weeks earlier and leaving earlier too. Plants, animals, insects, all are trying to cope with changing weather patterns brought about by climate change. Mother Nature is in charge more than ever before, even causing human residents to be more than a little nervous.

Strong, coastal winds have always been unpredictable. The wind tosses at will pleasure boats and fishing boats alike, such as the demolished shrimper heaped on the beach near the jetty not far from where I stand. No lives were lost, but some ships in the past weren't so lucky. Often the sea was calm, then suddenly churning with huge waves that toppled and smashed ships, their cargo, and crew.

The earliest visitors to this peninsula were tribes of the coast and those from the Chehalis Valley inland who came to dig clams and trade, prior to white man's arrival. The first white man, Captain Robert Gray, anchored in the bay on May 7, 1792. With sandy beaches and forested interior, this peninsula must have looked like heaven after many months at sea.

Over 150 years later it still looked like paradise to throngs of retired folks and city dwellers seeking a quiet retreat by the sea. Seeing prime real estate for vacation homes and retiree bungalows, investors began selling lots. Soon homes sprouted on the flat peninsula, spawning the birth of a city in 1970.

Jutting into the Pacific, the Ocean Shores peninsula is right in the path of tsunamis caused by shifts in the offshore Cascadia subduction zone as well as land-based earthquakes. Even with some warning, it would be nearly impossible for the 5,500 residents to evacuate quickly enough on the one two-lane road connecting it to the mainland.

Federal and state agencies, along with University of Washington researchers and an engineering firm, began planning the construction of a vertical refuge called Project Safe Haven in the town of Ocean Shores, designed to hold townspeople as well as visitors. Plans were also developed for vertical refuges in the coastal towns of Westport and Long Beach. To date, only the town of Westport has a completed structure, incorporating it into a new elementary school.

The vertical refuge might save lives in the event of a tsunami, but it's designed to be a temporary sanctuary. Where will residents go when the warming, expanding ocean causes the sea level to reach farther and farther up the beach permanently? Even with rock walls and other fortifications already in place, the west side of the peninsula, as well as the entire west coast of the state, will likely be one of the first areas to experience sea-level rise.

My thoughts wander back to a visit to Ocean Shores the previous spring. The chilly weather with threatening skies had kept most visitors home. From the parking lot at the jetty, it's a short walk to the beach, where waves throw themselves upon the shore. I was the only one on the beach that morning as I meandered along the sand, dodging the foam creeping closer until it was just inches away from my faded navy Keds.

Walking back to the parking lot, I noticed something skittering across the sand. Among strings of seaweed, bits of weathered driftwood, and broken shells, a tiny sandpiper stood its ground about ten feet away, staring intently at me as I slowly passed by.

"This is my territory," his eyes seemed to say. "I came a long way and still have a long way to go. Just let me rest."

The wind that would take him to his breeding grounds in Alaska ruffled the feathers on one wing as he stared at me. Looking at the tiny, vulnerable shorebird, I hoped that this beach still provides a rest and refuel stop next time he drops by on his long journey.

MARINA BEACH

This April morning is unseasonably warm. From my perch on a log, I watch a sleek white ferry glide across Puget Sound toward Whidbey Island. A few young children explore the water's edge with their attentive adults trailing behind.

When Lieutenant Charles Wilkes, USN, landed at this beach in 1841,* it was just a narrow, sandy shore with towering evergreens creeping westward toward the Pacific. A promising location for exploration, he named the spot Point Edmund, after his son.

Sixty years later, in the early 1900s, a shingle-manufacturing mill sprawled along the shore with rail lines leading north and south. Cedar trees from the lush forests behind the mill provided a steady supply of shingle material.

Today the landscape where the forests grew has become the town of Edmonds, population forty thousand, with a spider's web of roads and buildings sprouting everywhere. At the end of a walkway along the waterfront, past restaurants and the marina, a gravel-and-sand beach hugs the shoreline at Marina Beach Park.

About midmorning on warm days, young mothers with toddlers drift in, staking out picnic spots among the driftwood logs. Soon plastic buckets, shovels, and toy trucks, along with picnic coolers, mark the families' spaces on this sandy turf.

Near the parking lot, a small curved driftwood log has been painted brown with a grotesque red snout. Someone thought it

---

* "Wilkes, Charles: (1798–1877)," HistoryLink.org, accessed July 28, 2018, http://www.historylink.org/File/5226.

resembled a whale. But the whales that sometimes swim close to this shore are not brown.

Black-and-white orcas can sometimes be seen between Edmonds and Whidbey Island, never very close to shore. Now an endangered species, the orcas have fallen victim to pollution of their watery habitat and our insatiable appetite for Chinook salmon, their favorite food.

Overfishing, nitrogen and phosphorus runoff from agriculture, urban runoff, and chemical pollution have wreaked havoc on our waterways. Some salmon returning up their natal rivers to spawn are dying before they reach their breeding place due to increased temperature of the water.

Gray whales migrate through these waters spring and fall between Baja, California, and Alaska. What a thrill it was to watch a mother gray whale and her calf near the fishing pier one April morning a couple of years ago. A gray hump broke through the surface of the water, then disappeared below to the delight of about a dozen onlookers. Even these whales may soon find the supply of marine organisms they feed on diminished due to warming waters. Climate change is disrupting many ecosystems, none more than the ocean.

I recall a talk on climate change I attended several weeks earlier. According to Shallin Busch, PhD, a research ecologist with the National Oceanic and Atmospheric Administration, climate warming will cause a change in ocean circulation and stratification, making it more difficult for marine organisms to find their food. Effects on our fishing industry are already appearing.

Maybe it all comes down to too many people, wanting too much for themselves. They have forgotten about the creatures who share this planet with us. More people clogging the roadways, more people to be fed, more people contributing to urban sprawl, and it's accelerating the rate of climate change. The United Nations predicts that by 2050 the world's population could explode to 9.8 billion. Reining in population growth seems a logical step to help stop irreversible damage to our planet before it's too late. It has to start with each person.

In his book, *Sparing Nature*, Jeffrey K. McKee, associate professor of anthropology at Ohio State University, claims the way to slow population growth isn't difficult to figure out.

"The alternatives are so clear," he says, "a better life for fewer people or greater strife for more people. Conservation of the living world and taking responsibility for our reproductive habits."

According to McKee, there are two hundred thousand new people born each day. He claims that we will need twice as much agricultural land as we have now to feed a world population of nearly ten billion in 2050, as predicted by the United Nations.

Changing the expectations of society regarding whether or not to have children will play a significant role in making the transition to families having no more than two, or even zero, offspring. Tax breaks for couples with fewer children might be an incentive, he suggests.

Back at Marina Beach, a young pregnant mother carries one child along the water's edge as two others trail behind her. Isn't three enough? Maybe she wasn't thinking about the impact these children will have on resources in the future. Or, maybe they aren't all hers.

QUILEUTE COUNTRY

Clinging to a sliver of land on the western edge of Washington state, this beach feels remote, isolated. Driftwood is stacked helter-skelter by thundering waves. Treading slowly among the logs, you can find treasures hidden from human eyes for eons; I did on my last visit....

It fits my hand perfectly, this small smooth gray rock with a thin edge on one side. Was it a tool, perhaps a scraper for marine mammal skins, shaped by human hands hundreds of years ago? Members of the Quileute Tribe hunted marine mammals in the 1800s. It almost felt warm when I picked it up. This is not just any old rock, but one with a past. It looks like scrapers crafted by Native Americans that I've seen in museums. Discarded here on the beach, now it's pounded by the surf and trampled by the boots and tennis shoes of beachcombers exploring this narrow strip of land at the northwestern edge of Olympic National Park.

Historically a small tribe, today there are about four hundred members living on the reservation, with 750 enrolled. Their reservation is only one square mile, bordered on one side by the Pacific and the national park on the other three sides.

I rub the stone tool gently between my hands. I can't seem to put it down. My breath catches in my throat. It feels as if I have found something I lost long ago. A link with the past. My past? Should I take it home—maybe put it in a potted plant or on the kitchen window sill?

A gull perched atop a boulder watches me, then shrieks as if to say, "Put it down!"

He's right. It belongs here—bathed in the salty air with the other rocks, weathered driftwood, and, perhaps, the spirits of departed souls.

I wonder if those souls will be watching as their tribe builds a new village on higher ground to the east, away from the threat of tsunamis and the rising tides caused by a warming ocean. The Quileutes' contribution to climate change is negligible, but they are directly affected by the contributions of other people, other countries.

They watch in frustration as seawaters creep higher and higher onto the beach, flooding buildings near the shore. Only ten to fifteen feet above sea level, high tide brings seawater sloshing through parts of the town. More frequent heavy rainfall washes out roads.

No seals or whales are hunted from this beach now. Their ancestors hunted marine mammals, especially seals, from long canoes. None of the carved house posts or big potlatch houses remain that once lined the beach, but remnants still poke out of the soil in some places. In addition to sturdy canoes and house posts, the abundant western red cedar trees were put to good use for many everyday items. The soft inner bark was used to make capes, skirts, and conical-shaped hats that protected them from the 115 inches of annual rainfall. These early Native Americans took advantage of the natural bounty of their homeland to live a comfortable life, long before outsiders came along and saw the acres of valuable timber ripe for the taking.

A treaty in 1889 established the reservation on over eight hundred thousand acres of timberland on the heavily forested western slope of the Cascade Mountain Range as well as offshore waters. The treaty also guaranteed the hunting, fishing, and gathering rights to which they were accustomed. They were officially recognized as a tribe in 1936.

These days "Move to Higher Ground" (MTHG) is their motto as they make plans to move the village uphill onto 772 acres of state land that previously belonged to the tribe. In 2012 President Obama signed a law transferring the land back to the tribe.

I drop the stone tool between two driftwood logs as I head back to the parking lot. Though I left the scraper on the beach, I'll keep the image of it in my mind and the feeling of the smooth stone in my hand long after it has settled on the sandy bottom of the rising seawaters.

SACRED GROUND

"When you strip bark, you take it from the butt end of the tree up, and when you get it off the tree, you roll it up," explained Fran James, elder of the Lummi Tribe, whose heritage has been embedded in a bountiful corner of sea and land in northwestern Washington for thousands of years. I marveled at the skillful manner in which her sun-tanned hands rolled the rough bark around a stick. Those same hands helped other tribal members of all ages learn to weave the bark into traditional Lummi baskets.

Fran spearheaded the revival of Coast Salish wool weaving; both her blankets and her baskets are in museums and private collections worldwide. When she passed away on April 28, 2013, she was a master basket weaver who had shared her basketry skills with hundreds of tribal members, especially youth, as well as people outside the tribe. At the same time she was teaching the youth basketry, she wove in important values. "All these little pieces, we use them for tying," she continued. "We don't waste anything."

That spring day years ago was cool, the sun ducking behind clouds off and on as we wandered through a forest on reservation land. My late photographer husband, Jim, his friend Howard Schuman, who worked in television and public relations, and I were working on a project to create a video of Indian basket makers of Washington state while some of the master basket makers were still alive.

Fran's son, Bill James, accompanied us. In response to Howard's question about why they make baskets, Bill thought a moment, then explained, "We make them to preserve the traditional art. Teaching young people of the tribe to carry on this tradition gives them a sense of self-identity and accomplishment." Later that day in the basketry classroom, I watched Bill's teacher skills take over as he confidently guided several young students making their first baskets.

Passing along traditions to the youth so they can continue to live in harmony with the environment, while they gather the abundance around them, has been the Lummi lifeway for 3,500 years. Everything they needed came from the sea and the land. Harvesting salmon from the rivers and the Salish Sea continues to be important to the tribe both economically and spiritually. As salmon runs have dwindled over the years, the Lummi were among the first to begin working on riverbanks that had eroded due to poor logging practices.

Presently a Lummi Nation hereditary chief, Bill James is intent on seeing tribal lifeways continue for generations to come. He became a vocal activist upon learning about the potential impact of a proposed coal export terminal at Cherry Point near Bellingham in 2013, at the very site of ancient tribal fishing grounds that still support a significant fishing industry for the Lummi. The ancestral tribal fishing grounds there are so much a part of their culture, he commented, "Saving it [Cherry Point] preserves the tribe's very way of life." An ancestral burial ground also lies beneath the sand and gravel at the site.

Cherry Point has been a sacred fishing grounds for many generations. An oil refinery and an aluminum smelter already claim spots in an existing industrial facility nearby. Shipping lanes serving these two facilities already reduce access to the fishing grounds, while the ships often destroy Lummi crabbing gear.

In 2011 Peabody Coal and SSA Marine filed a request with the Army Corps of Engineers to build a deep-water port at Cherry Point to support a new coal export terminal capable of shipping forty-eight million tons of coal per year. Up to eighteen additional coal trains per day would travel from Wyoming's Powder River Basin, down the Columbia River, then north along the waterfronts of Seattle and towns to the north, to be offloaded onto massive vessels the size of three football fields and heading for China and India. According to the Power Past Coal initiative of the Sierra Club, a few days after being burned in these countries, toxic coal emissions would travel back to the West Coast on the jet stream.

I was among the nearly one thousand red-shirted "No Coal Exports" activists at a hearing on the Cherry Point Coal Export Terminal held in Seattle in December 2014. Though speakers came from all walks of life, only a handful of them were supporting the project. It was exciting to hear a similar message delivered by tribal members and ranchers from Montana, where the coal would be mined, and others along the route of the 120-car open coal trains that would snake their way across the state to the proposed export terminal.

In April of 2016 the permit was denied. However, with the current presidential administration determined to revitalize the dying fossil-fuel industry, that permit may be approved. What a travesty that would be. The Lummi's treaty rights matter. Salmon, crabs, herring—they all matter. Our warming climate is approaching the point of no return. If we are to avoid catastrophic changes to our environment, the time to take action is now.

Winter Visitors

SITE 28: EDMONDS NORTH
START TIME: 1335; END TIME: 1400
WEATHER: OVERCAST
SEA CONDITIONS: CALM

"Two red-necked grebes to the right!" shouts Beth. Just as I find them with my binoculars, they dive. "They were at twenty degrees," says Kay as she reads the numbers on the compass attached to the top of her spotting scope. "Forty millimeters," (from our measuring point on the far horizon) I add, squinting at the tiny "mm" numbers on the ruler I hold at arm's length in front of me. I quickly pull up my fleece scarf against the biting wind this freezing January morning, then refocus my binoculars. Standing out on the fishing pier, we are directly in the path of that angry wind.

Barbara quickly jots down the numbers on a recording form. A few days after the survey, she will email to me the entire list of species we observed at our two sites along the Edmonds waterfront so that I can enter the information into the online database for Seattle Audubon's winter Puget Sound Seabird Survey.

Our team of four observers has volunteered with the survey since it began in 2007. The first Saturday of each month, October through April, for a period of not less than fifteen minutes or more than thirty minutes, sometime during the four-hour window, two hours on either side of high tide, we bundle up, grab our gear, and head toward the fishing pier, Site 28. When we're done there, we go south about a quarter mile to Site 27, just south of the fence around the dog park at Marina Beach.

At times, near-gale-force winds have nearly blown recording forms into the water; rain has streaked our binoculars; or blinding sun glaring off the water has made it almost impossible to tell a loon from a cormorant. My frigid fingers often turn blue inside my gloves before we're finished.

The survey was developed to gather baseline data on winter seabirds along the central Puget Sound coast and Strait of Juan de Fuca. The 2016–2017 season had 121 survey sites covering an area that includes 2,400 acres of nearshore saltwater habitat from Olympia north to Deception Pass, a strait separating Whidbey and Fidalgo Islands, as well as sites on the Olympic Peninsula. All volunteers are trained in the same scientific protocol so results are consistent and reliable. Beyond gathering baseline data, the information our four-person team and the other nearly two hundred citizen-science volunteers are gathering will be useful to state agencies in the event of an oil spill as well as to those studying climate change.

It's rewarding to know that I'm helping by gathering this data, but I'm also personally interested in finding out which species are decreasing or increasing in number. Some years the count is way down for a certain species, or a newcomer drops by, like the brown pelican that surprised us two years ago. A compilation of data from the past eight survey seasons indicates four species in our survey area are declining in numbers, including black brant, one of my favorites. According to information from National Audubon on seabirds nationwide, 47 percent of the seabirds are decreasing in number, 25 percent are stable, 17 percent are increasing, 11 percent are unknown.

Small grants funded our seabird survey in the beginning. In 2014 additional funding came through from the Environmental Protection Agency's National Estuary Program via the Washington Department of Fish and Wildlife's Puget Sound Marine and Nearshore Grant Program to expand the project over to the Olympic Peninsula, where there is a greater likelihood of an oil spill.

"Oh, look!" exclaims Kay. "Brant! That whole group way out there. Hmmm, I count about . . . seventy or eighty." The smallest and fastest fliers of all geese, these striking black-and-white sea geese pass along the Edmonds waterfront each spring and fall.

While some of these Pacific black brant winter along the British Columbia, California, Oregon, and Washington coasts, 90 percent migrate to Baja, California. Those Baja travelers make the 3,400-mile journey nonstop in fifty hours. Other brant, the western high Arctic brant, with gray bellies, also winter along the Washington coast, primarily in Skagit and Whatcom Counties in north Puget Sound.

Over 140,000 Pacific black brant migrate along the Pacific flyway each year, with most returning to the coastal tundra of the high Arctic to breed. The brant that winter in Baja begin heading back north in mid-February, stopping in estuaries along the way to feed and rest. I can't help but wonder how they make it back year after year when more and more of their feeding stops are becoming strip malls or motel parking lots.

By late April, they reach Izembek Lagoon in Alaska, where they are joined by those wintering along the Washington coast, remaining for a month before departing for their nesting grounds on the Arctic tundra. Their diet consists mainly of eelgrass, which grows in intertidal mudflats or along the edge of the shoreline. Brant usually lay three to five eggs in a ground nest lined with down the female has plucked from her breast.

Chicks grow quickly on a diet of protein-rich insects. With spring now arriving a week or two earlier each year, timing their breeding to coincide with the abundance of insects to feed their chicks may be tricky. If the insects begin breeding earlier due to the warming climate, I hope the brant can reset their biological clocks to keep up with their food source.

After breeding, the brant fly to Teshekpuk Lake in the Arctic to molt. There they feed on nutritious sedges that fuel the production of new feathers. When the temperature drops and a low-pressure system moves in, the brant know it's time to head south.

According to a U.S. Geological Survey researcher at the Alaska Science Center, global warming seems to be driving higher rates of coastline erosion and higher storm surges that make Arctic lakes salty, robbing them of the freshwater habitat that sedges need to grow. The warming climate has already had an effect on migratory habits for some who now remain along the Alaska Peninsula throughout the year.

The population wintering in Skagit and Whatcom Counties in Washington is closely monitored by the Washington Department of Fish and Wildlife (WDFW). "For the past several years, we have had at least six thousand birds in the winter when we conducted our survey," explained Christopher Danilson, wildlife biologist with WDFW, in a recent e-mail exchange. "We definitely see minor oscillations from year to year. Last year we were just shy of nine thousand in Skagit bays, which was up from sixty-seven hundred the year before. However, winter weather can affect these surveys and even a change of a couple thousand birds can be reflective of a single breeding season, change in migration, or winter distribution," he continued, adding that if the numbers are too low, the hunting season is shortened.

While I realize that the agency is partially funded by hunting licenses, it seems morally unjust to allow hunting of these diminutive, elegant geese that are already coping with a myriad of challenges, such as habitat destruction and the effects of global warming on their wintering and breeding grounds.

Back at the fishing pier, we continue to add seabird sightings to the recording form. "Five goldeneyes just beyond the rock jetty to our left," Barbara calls out over the horn blast of a ferry approaching the Edmonds ferry dock. The others focus on the goldeneyes, trying to figure out if they are common or Barrow's. But I'm still focused on the brant, hoping we can continue to add this beautiful visitor to our winter seabird survey for many years to come.

TRAVEL DISCOVERIES

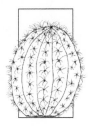

Eye Contact

An elegant young male pronghorn, white coat dashed with designer markings of caramel color, approached the road edge between our Yellowstone Park bus and a tourist vehicle in front of us. Behind the pronghorn, a dozen more grazed contentedly. He gazed across the road to tempting feeding grounds on the other side, then glanced at me standing alongside the bus. The poor guy looked confused and frustrated. Perhaps he felt that it was not safe to pass between the car ahead of us and our small bus. A man stood beside the car, taking photos. Our tour leader motioned for the driver of the car to move along.

In my imagination, I stroked the smooth, creamy fur on his back, telling him it was okay now. Gathering his courage, the pronghorn crossed the road, then broke into a sprint.

To make eye contact with a wild animal is like reaching into another dimension. Strong, confident, adaptable, that animal has learned to survive in a world I can't even imagine.

I'm not talking about making eye contact with a squirrel gathering acorns that have fallen on the sidewalk as I walk to the gym. Or about the defensive crow swooping within inches of my head, glaring at me squawking, "Get away from my nest!"

I'm talking about the wilder critters—bison, elk, bears, bighorn sheep, pronghorn antelope, wolves, and others that live in places like Yellowstone National Park. Thousands of people visit that park every day, most during the summer months. Aren't all those vehicles with their gawking, intruding humans having an impact on the animals? I wanted to find out for myself.

For my first trip to Yellowstone, I enrolled in a Yellowstone Institute three-day field seminar, "Autumn Wildlife Watching." And watch wildlife we did, lots of wildlife. So did scores of other people. Lines of cars pulled over to the side of the road in "bear jams" were long, but not the miles long that they had been for summer visitors.

Visitation has increased 20 percent in the last two years. By October 2017, over four million visitors had descended on Yellowstone. With the number of tour groups, especially from outside the United States, increasing, that number will continue to rise. "Even with all these people, it's hard to believe that 98 percent of the park still belongs to the animals," commented our guide.

According to Park Superintendent Dan Wenk, 2016 tested the ability of the park to handle that many visitors, resulting in traffic jams miles long. Not surprising that, with more visitors, there were more human/wildlife encounters for rangers to resolve. Park regulations state that people must stay twenty-five yards from bison and seven hundred feet from a grizzly. Since 2011 there have been three fatal grizzly attacks. These usually result in the grizzly being killed so it will not attack again.

Dawn was breaking when we headed out at 6:30 A.M. each day. As the curtain of night lifted, the view out our bus windows was a panorama of beige dried grasses, dark pines punctuated with vivid-yellow quaking aspen and cottonwood groves. Bright gold rabbitbrush and other shrubs lit up the understory. The smoothly undulating hills and valleys belied the violent volcanic birth of Yellowstone. But even the smoldering magma beneath our feet couldn't coax the morning temperature above forty degrees.

Yellowstone was alive with birds. Clark's nutcrackers hung around pine groves along with pine siskins. Red-tailed hawks and northern harriers rode the afternoon thermals while a few osprey snatched trout from lakes we passed. Bald eagles seemed abundant.

Other than getting to see the lay of the land and some of the thermal features, highlights for me were making eye contact with a few species, like that pronghorn.

There are about three hundred to five hundred pronghorn antelope in the park. According to park scientists, there should be about a thousand to maintain a viable population. These graceful athletes can run at speeds up to fifty-five miles per hour, but the track is not always clear since migration routes are often constricted by fences, highways, subdivisions, and oil fields. Surviving members of a species that evolved twenty million years ago, they could face extirpation from a severe winter or disease since the population is so small.

It felt like at least eighty degrees one afternoon we took a short hike to Lost Lake. Brisk warm breezes pushed us along the narrow upslope covered in shrubby plants with a road at the top on our left.

Rounding a bend, we stopped dead in our tracks. Lying about twenty feet off the trail on our right side was a bison. His small bloodshot eyes turned our way, but he didn't move. I wanted to walk over to him and touch the dark-brown tightly curled fur on his head. Would it feel coarse, like a sheep?

We scrambled up the slope on our left, continuing toward the lake. Once he was out of sight, we descended the slope to the trail and continued. Rounding another bend, we encountered another bison, this one a little farther away from the trail, resting in the shade, his long black tail swishing flies away. Bison can run thirty-five miles per hour but have a hard time running uphill, so we did have a way out if he decided to charge. But at that moment, he was busy munching grass, so we continued on the trail. After a look at deep-cobalt Lost Lake, we headed back down along the road on top of the left side of the trail, giving the bison more space. The two bison near the trail, along with six feeding farther back near some trees, didn't seem concerned about our presence.

There are about five thousand bison in the park, though the ecosystem can sustain only three thousand. At the onset of winter, bison head down to the lower elevations, where a thinner snowpack and less brutal temperatures make it easier to find food. But they face other obstacles as well. Spreading agricultural fields and burgeoning development crop up where they used to graze.

Once they leave the park, they are fair game to hunters. Thousands of Yellowstone bison have been slaughtered because some ranchers are concerned they will pass brucellosis, a bacterial disease, to their cattle. Cattle grazing on public lands can routinely come in contact with bison; however, in the past thirty years there have not been any documented cases of bison transmitting the disease to cattle.

The next morning, we pulled off to the side of Tower-Canyon Road where other cars were lined up. Cameras clicked as two seven- or eight-year-old bighorn sheep fed on thistles in a ditch near the road. Many of these wildlife watchers were closer than the twenty-five-foot minimum space allowed for viewing these animals.

The bighorns turned and headed south along the ditch, toward the few of us standing on the grass at the end of the ditch. Heads lowered, shoulder to shoulder with eyes focused on these intruders, they swaggered slowly toward us, appearing like tough guys belonging to the local Yellowstone bighorn gang. We began retreating to our bus....

As of 2014, there were about 420 Rocky Mountain bighorn sheep in Yellowstone. Living in small, fragmented groups, they are vulnerable to disease, habitat loss, and disruption of their migratory routes by encroaching development.

Throughout the three-day field seminar, we occasionally spotted black bears gorging in preparation for hibernation. Several black bears upslope or downslope of the road were feeding on cones from mountain whitebark pines. The bears crunch the whole cone, but get nutrition only from the tiny seeds inside.

As a result of global warming, mountain pine beetles have destroyed thousands of acres of mountain whitebark pine trees. Warmer temperatures are causing the beetles to invade higher elevations, where the tree grows. The cones are also a food source for the seven hundred grizzlies in the park. Listed as an endangered species for decades, the grizzly was delisted in 2017, mainly due to pressure from hunters.

Climate change is altering many natural cycles in the park. Warmer temperatures mean a longer growing season with some plants blooming earlier and pollinators arriving earlier. Huge, destructive forest fires are the result of a fire season beginning earlier and lasting longer.

Besides a warming climate altering their food source, animals have to deal with an increasing number of tourists invading their habitat. This must be having an effect on the animals.

The elk herd that frequently hangs out at the Mammoth Hot Springs Hotel provides a good example of tourism having an impact on the wildlife the park has a mandate to protect. Drawn by green lawns for feeding and resting, the herd takes over in the late afternoon, much to the delight of visitors. Elk feed on the tender grass shoots, nurse young, even mate while cameras catch every moment. A study done on elk to measure blood pressure levels indicated elevated blood pressure when people were nearby.

It's the paradox of the "cultivated wild." Manage the park for wildlife or for tourists?

I agree with some of the scientists and officials governing Yellowstone that the present level of tourism is not sustainable if the resources that the park was established to protect are to be preserved.

Superintendent Wenk explains that the park service must decide whether to allow unrestricted access to the park, knowing that the resources will eventually be damaged, or restrict access to protect the resource.

Our guide, who has been with the park for thirty years, was more adamant. "The park can become a zoo in the summer with bear jams, vehicles stopped alongside the road for miles. But, officials won't restrict the number of visitors until something awful happens and they are forced to," he continues. "It's the locals; they don't want to lose any business."

Seems like a no-brainer to me. Officials need to take direction from reports generated by park scientists instead of tourists and businesses concerned about possible lost revenue. Yosemite and Denali National Parks both have restrictions on the number of visitors to protect the resource. Yellowstone needs to follow suit.

FUR SEALS OF ST. PAUL

Gale-force winds pushed me down the path to Gorbatch northern fur seal rookery on the island of Saint Paul in the Pribilofs that misty August evening in 1982. The pups' loud "awk! awk!" cut the chilly air as they awkwardly pulled themselves across the boulders, trying to stay out of the way of the huge, ruthless bulls. Knowing the eventual fate of some of these seals made the scene feel raw, even dismal, to me.

At forty square miles, Saint Paul is the largest of the five Pribilof Islands, which lie three hundred miles from mainland Alaska in the Bering Sea. Blustery, foggy days are the norm while sunny days are rare, even in summer.

Camera always at the ready, I spotted a dark, odd-shaped object near the side of the dirt road a short distance ahead of me. Getting closer, I could see that it was the severed flipper of a fur seal. The fields surrounding me, now lush with wildflowers, only weeks ago were the bloody killing fields for the annual slaughter of twenty-seven thousand northern fur seals.

The 480 Aleuts living on the island today are descendants of those brought to Saint Paul and Saint George Islands several hundred years ago by Russian settlers. At that time, the market for sea otter pelts was lucrative; Aleuts were good hunters. When overhunting wiped out the sea otters, the hunters switched to the fur seals. The seals were nearly hunted to extinction until commercial seal hunting on Saint Paul was discontinued in 1984.

Saint Paul has the largest northern fur seal rookery in the world, with several hundred thousand coming ashore to breed in early June each year. Around 1,500 are still killed for subsistence, but fewer are consumed each year due to other food options now offered in the island market.

Many Aleuts on Saint Paul claim the seal hunt is an integral part of their cultural identity. But times have changed. Life is more challenging for many marine mammals because of dwindling food sources, including the declining seal populations. As younger Aleuts embrace a more modern lifestyle, including food choices, even subsistence hunting of seals may die out.

It's the same argument made by the Makah living on the northwest tip of Washington state who claim they need to hunt gray whales to maintain their cultural identity. Overfishing along with warming ocean waters are disrupting the food supply for many of these marine mammals, causing stress and contributing to population decline.

According to Rod Towell, a statistician for the Department of Commerce's National Marine Mammal Laboratory, northern fur seal numbers were way down in the 2016 survey, possibly due to diminished food supply or disease or other unknown factors.

Growing up in an environment where slaughtering animals reinforces an aspect of the culture is bound to shape the values of the young people. When I visited Saint Paul, it was not uncommon to observe boys on ATVs harassing the seals. They would drive as close to the rookery as possible, then race their engines to frighten the seal pups. Some of this disregard for the suffering of other living creatures frequently affects other areas of their lives. Research done by the Humane Society of the United States, the FBI, and other agencies has revealed that becoming insensitive to the suffering of animals can often promote insensitivity or cruelty toward humans.

Shortly after the commercial seal hunt ended, the U.S. government supplied $12 million to develop and diversify a fishery-related economy. This now includes a small boat harbor with a vessel repair facility, seafood processing plants, as well as an expanded fishing fleet. The improved economy supports additional educational and job training opportunities along with upgraded living conditions.

With the booming fishing industry come risks. The Pribilofs are surrounded by the richest bottom fishery in the world. As many as three hundred fishing vessels at a time ply the waters, increasing the risk of oil spills. Several years ago a large spill caused major damage to the marine ecosystem, including sea-duck habitat.

An improved economy is slowly propelling them into the twenty-first century. Since 2007, there have been three wind turbines along with a new electrical generation facility installed on the island. Around seven hundred ecotourists visit the island each year to view the seal colonies and cliffs where millions of seabirds jostle for a spot on the rocks to lay their eggs.

I didn't take a photograph of the fur seal flipper that evening, but have not forgotten it either. Hopefully, the Saint Paul islanders will continue to improve their living conditions, embracing the twenty-first century, until fur seal hunting is alive only in their stories and not on the bloody killing fields.

Inhabitants of the Heath

Intoxicated by the delicious nectar at the bottom of the pitcher plant, the fly's threadlike legs inch downward toward his pot of gold. But instead of nectar, a pool of digestive enzymes awaits. His body begins to dissolve as he enters the pool, two legs desperately reaching for a way out. But the downward-pointing hairs lining the inside walls create a trap from which there is no escape. Shortly, all that remains of the fly is a shiny black smudge of melted wings, legs, thorax, and head. Click!

Cleaning out some of the magazines where my articles have been published, I recently came across this one featuring my photo of a pitcher plant. On a cold, rainy Northwest morning, it was enjoyable to access memories of that warm, muggy bog. Details from that trip jotted in tattered notebooks made it easy to recall the experience.

Pitcher plants and other unusual wetland fauna are just one feature of Acadia National Park, thirty-three thousand acres of bays, islands, coastal tide pools, estuaries, and forests, clinging to the midcoast region of Maine. Over 20 percent of the park is wetland. Aside from providing habitat for unique plants, amphibians, and other animals, wetlands play an important role in filtering pollutants out of storm-water runoff and buffering the land from coastal storm surges.

Acadia's wetlands and other diverse ecosystems have remained relatively stable over the past one hundred years since the park was created. But with climate change running rampant, I wondered how it was affecting the park that I visited and photographed about twenty-five years ago.

Looking back at that visit, I now realize that the ecological role wetlands play will become even more important as the climate warms and weather patterns change. Faced with budget and staffing cuts for years, park management now has to come up with strategies for dealing with a growing number of visitors along with the impacts of climate change.

On that hot, humid day I recall stepping gingerly around the mushy edges of the Big Heath, one of the wetlands I explored, and squatting unsteadily on the uneven ground to focus my Nikon lens on the unusual plants that were thriving in this mucky environment. I photographed carnivorous pitcher plants and sundews as well as gorgeous orchids, batting away annoying black flies all the while.

After a bit, the muggy air made me feel slightly woozy. Looking through the camera lens, I saw plants, water, and trees blending softly as if in an impressionist painting. The spruce and tamarack ringing the edges of the heath became a green swipe behind brownish sphagnum moss and sedges that formed a spongy mat in the center. Water lilies floated on top of the muck. A green frog squatted on a lily pad, waiting patiently for the slightest movement signaling the approach of a dragonfly.

Warmed by the summer sun, pink and yellow orchids sprouted from the brown water. Not far from my knee, a mouse hugged the soggy earth under low cranberry bushes, seeking shelter from a hawk overhead as a Lincoln's sparrow scrounged in the duff for seeds. Footprints left by deer, muskrat, and beaver filled with water around the edge of the quagmire. Silence prevailed, broken only by the occasional buzz of mosquitos and black flies. As summer ends, the white blossoms of Labrador tea would spice the air; a few weeks after that, the magic wand of autumn will cloak the landscape in crimson.

But things are changing. According to Abe Miller-Rushing, science coordinator for the park, invasives like purple loosestrife, glossy buckhorn, and barberry are encroaching on the wetlands. The native orchids that proliferated when I was there years ago are getting harder to find.

Within the next century, Maine's temperature is expected to increase four degrees with more precipitation and more severe storms. In an email, John Kelly, management assistant at Acadia, gave me more details of what was to come. According to Kelly, sea level has risen one foot since 1900 and is anticipated to rise an additional one to four feet by 2100. Flooding will likely occur in some picnic areas, beaches, trails, roads, and other low-lying areas during severe storms or high tides.

Spring events, such as flowering, leafing out, and bird migrations, now happen much earlier than in the past. Blueberries flower more than three weeks earlier than 160 years ago. These changes may be creating temporal mismatches between plants and pollinators, predators and prey, or other interacting species.

The park is already experiencing earlier spring thaws and heavier precipitation along with more invasive plants that are better adapted to the earlier spring thaws than some of the native plants. This is also increasing new insect pests and vector-borne diseases in the park, he adds.

Park scientists report that some of the most common trees, such as fir, spruce, aspen, and paper birch, and the birds that depend on them, may become less common. Oak, maple, and hickory may take their place. Warmer northern climates are forcing some species of plants and animals to move farther north while those from the south move north.

Near towns and downwind of industrial areas to the west and south, Acadia is in the path of winds carrying emissions from coal-burning power plants in the Midwest. Air pollutants such as ozone, particulates, and mercury are responsible for the hazy views visitors often encounter as well as damage to plants as these chemicals work their way into the ecosystem.

Looking at photographs from that visit, I am reminded that the fate of special places like Acadia is in our hands. We can start using clean energy and slow down the rate of global warming. Like the fly going after his pot of nectar in the pitcher plant, fossil-fuel advocates throw caution to the wind, casting aside protective regulations to reach their own greedy pot of gold. It comes down to choices. We can show our children and grandchildren pictures of the unique fragile environments that are gone because we didn't fight hard enough to protect them, or we can push harder to save them now. The choice is ours.

JOURNEY INTERRUPTED

6:00 P.M. APPROACHING THE TOWN OF WILLOW, ALASKA

Chugging along toward Anchorage on a sunny summer day, all of a sudden I feel a strong jolt, smell burning rubber, and hear the sound of screeching brakes as our Denali Star train comes to a halt. "Must have hit a bear," comments the guy across the aisle. White papers flutter past my window like feathers of a dove against the blue sky. A bear carrying papers?

For the last hour, we'd been treated to a picture-postcard view of snowy Denali looming above the birches and aspens. When I think of Alaska, where I lived a few years as a child, I remember snow, lots of snow. I remember walking to the school-bus stop in the dark, then walking home from it, still in the dark, in the winter. Sometimes the northern lights spanned the sky overhead, giving it a greenish glow. I don't remember ever being cold, but I do remember the warm, safe, and secure feeing I had living with my mother and sister and grandparents in their home. Grandpa and I were very close; I think I've really come back hoping to find him, though he died long ago.

We're scheduled to arrive in Anchorage at 8:00 P.M. Wonder why we're just sitting here? This seven-hour train ride from Denali National Park back to Anchorage provided some time to reflect on the magnificent six-million-acre park I had just visited. Watching the grizzlies, caribou, and moose feeding, fleeing, and adapting to the circumstances reminded me how far we have strayed from living in tune with nature as they do.

Just two days ago, I watched a bull moose munching willows right alongside our tour bus, as cameras clicked all around me. Moose in the park have become very tolerant of people, but people haven't always been tolerant of moose around the towns. Deep in the memory of my childhood self was a skinny moose that wandered into our Anchorage neighborhood one winter. All he wanted was to browse on some bushes; what he got was a bullet to the head. How sad.

Moose wandering into Anchorage was rare in those days. Now there are around 1,500 moose snatching apples from backyard trees, snooping around porches, rutting in driveways, giving birth on front lawns. When winter snow begins to creep down the Chugach Mountains, moose head into town for the easy life.

Glancing at my watch, I feel frustration rise as the delay has begun to irritate me. I guess that's typical. Like most people, I want an explanation of events unfolding that may have interrupted my schedule. Evolution has thrust us far, far away from our beginnings. Gone is the ability to quickly adjust thoughts and behavior to the present moment without the need to try to control the outcome.

I was more flexible as a seven-year-old kid, sporting warm seal-skin mukluks on my feet, the last time I rode this train. My mother took us from our home in Anchorage to the Fur Rendezvous, a Native Alaskan winter celebration in Fairbanks, on a snowy February day. It was a very long day, an adventure for my sister and me. We didn't care at all when we would get back home—my mother was taking care of those details. Now I was lulled by a landscape without snow, when the screeching brakes jolted me back to real time, booting me out of the control seat.

Shortly after the impact of the train's abrupt stop, a young conductor speaking into a handheld radio passes my seat. Hurrying along behind him is a tall athletic-looking man in khakis and a long-sleeve blue shirt. "We're heading back; I've got a doctor with me," the conductor reports into the radio.

6:30 P.M. DELAYED IN WILLOW

A calm woman's voice comes over the loudspeaker: "I'm sorry, folks. We'll be here a while. We've hit a vehicle." A few of the passengers in my car slide out of their seats, cameras in hand, and walk back toward the vestibule linking the train cars where there is an open window. Anxious, not knowing what has happened, if anyone was hurt, or how long we would sit here, I can't seem to sit still but stand, watching the rear of the train car, waiting for answers. I pull out a PowerBar and start munching.

6:50 P.M. DRAMA IN WILLOW

About half an hour later, the doctor walks back up the aisle, a blank look on his face. I stare at him as he passes my seat; our eyes meet, but no expression crosses his face. Not a good sign.

10:30 P.M. STILL STUCK IN WILLOW

"We'll be leaving shortly, folks. Since we're not sure how much damage was done to this train, we're not taking you all the way to Anchorage," Miss Stay Calm announces. "We've arranged for motor coaches to be waiting for us three miles down the track at White's Crossing. We'll unload all the baggage on the ground, so go and retrieve yours before you board a motor coach. They'll take you to the train depot in Anchorage, where we've arranged for taxis to be waiting."

Unwrapping another PowerBar, I recall the image of my grandfather, his overalls covered in grease, coming home from his job as a boilermaker at the Alaska Railroad roundhouse in Anchorage. No matter how tired he was, he always greeted us with a big smile. If he were here now, maybe he could help fix this train. An old sourdough, Grandpa was a master at fixing things, adapting, whether it was panning for gold in the dead of winter or building an igloo out of snow blocks for my sister and me in our front yard. Now, having to adapt to an unexpected situation when I have my own agenda sometimes irks me.

Settling back in my seat, I melt into the memories of the past few days. Hiking in the park on the second day, I had watched an adult caribou with a juvenile trotting beside the road, a strategy to avoid bears and wolves, which are always on the lookout for juvenile caribou and less tolerant of vehicles. Each of the species in the park has adapted to its environment in order to survive. Our human species sometimes isn't willing to adapt to a situation, but instead tries to control it.

11:45 P.M. LEAVING WILLOW!

The train begins inching along. Restless passengers grumble to each other as they fiddle with cell phones and other gadgets, trying to update friends and family.

12:00 A.M. WHITE'S CROSSING

People line up, eager to get off the train. My window becomes like a television screen, broadcasting news of some far-off disaster. In the pitch darkness, a fire truck shines a spotlight over heaps of baggage unloaded onto a gravel lot. Red taillights from a row of motor coaches line the perimeter of the lot while about a dozen orange-vested disaster responders in yellow hard hats assist weary passengers off the train, help them locate luggage, and offer assistance carrying it to the waiting motor coaches.

"Welcome aboard, folks. We should be in Anchorage by 1:30 A.M.," the cheerful driver tells his busload of groggy riders. Our motor coach joins the caravan of seventeen others headed into the darkness toward Anchorage.

1:35 A.M. ANCHORAGE RAILROAD DEPOT

Baggage is unloaded onto the sidewalk as railroad personnel assist passengers to waiting Yellow Cabs or to phones inside the station, where they can call hotel shuttles or friends to pick them up.

Staring out the window of the Holiday Inn shuttle van, I balk at returning to these streets, these buildings, these schedules and focus instead on the animal encounters of several days ago.

It seems to me that we've severed our sensory awareness and a strong personal connection to our environment, weakening our ability to survive, relying instead on fleeting human connections to help us along on our journey.

Back in the park, I observed adaptive behavior firsthand. Cautiously, a female moose had sidled up behind a young alder, her eyes fixed on a grizzly in the distance. Watching and waiting, watching and waiting. She didn't begin feeding until the grizzly had disappeared. Her instincts informed her actions. How did we evolve to be so far apart, the moose and I?

Learning the reason for our delay when I read the *Anchorage Daily News* in the airport the next morning made my feelings of frustration seem so inconsequential. A woman learning to drive a stick-shift vehicle had somehow caused it to jump onto the tracks just in front of the oncoming train. I can take this train again sometime; she will never drive that car or see her family again. I guess it's all a matter of perspective.

Though we humans don't always adapt to changing circumstances as well as we could, I hope the animals in Denali National Park are able to adapt to a warmer, soggier home as the permafrost begins to melt, causing changes to their habitat. But, if they have adapted to coexisting with each other for so long, I'm confident most can cope with these new challenges as well. I sure hope so.

On the Trail of the Oreodont

A sizzling sun can send your mind spinning if you're not careful. I know—it happened to me. Exploring an oppressively hot rocky place one spring day, the past easily pulled me from the path I was following that afternoon. This happens sometimes; my imagination leaps out to experience another time, another place. What did it look like back then? What grew, what growled? I was about to find out....

After several short hikes in eastern Oregon's John Day Fossil Beds National Monument one day in May, I dropped into a sweaty heap under a shade tree on the grounds of the visitor center. Eyes closed, I sipped lukewarm water while trying to recover my energy. After a couple of hours in that oppressive heat, I was beginning to feel like a space cadet.

Step into my time machine, friends. Travel back with me, twenty-five million years ago to explore a world we can only imagine. We're in a hot, steamy subtropical forest on a chunk of land that will become part of central Oregon.

Listen. Do you hear something splashing and bellowing over there in the swamp? That herd of small, brown animals munching on swamp grass, those are oreodonts. Sort of look like sheep, don't they? (Oreodonts were mammals closely related to pigs, with no other close relatives living today.)

See those volcanoes in the distance—gushing lava and ash? Smell the sulfur fumes? These subtropical forests, swamps, the oreodonts, and other creatures living here will soon be buried under layers of ash and mud from the volcanoes. Their bodies will be preserved for people to study millions of years from now. Ash already has buried huge browsing animals like brontotheres and amynodonts that flourished eons ago right where we are standing.

Wait! Look up in that tree! That saber-toothed cat is eyeing that oreodont! He's jumping onto the oreodont's back and . . .

Fast forward to fifteen million years ago. Lava flows only intermittently now. There are moderate temperatures, a drier climate, and fertile volcanic soil. Hardwood forests are home to horses and deer that will remain in the John Day Valley for centuries. The bear dogs, rhinos, and camels will disappear.

Zip forward in time to eight million years ago. The climate has become drier and cooler. Savannahs are home to horses, pronghorns, and peccaries. As the land dries out, erosion begins.

Take a giant leap forward in time to spring 1994. Yellow, pink, red, and lavender wildflowers decorate the rolling hills of this desiccated landscape.

Established in 1975, this fourteen-thousand-acre national monument contains some of the richest fossil beds in the world. It is divided into three sections: the Clarno Unit, the Painted Hills Unit, and the Sheep Rock Unit.

Marked by the cliffs of the Clarno Palisades, fossil beds in the Clarno Unit have revealed remains of prehistoric seeds, nuts, fruits, leaves, and roots. Sandstone hills draped in vivid layers of red, pink, bronze, tan, and black minerals attract photographers to the Painted Hills Unit.

Casts of jawbones, tortoise shells, and other fossils line the walls of the visitor center, along with posters of the animals that once roamed where the building now stands. Though the prehistoric animals are long gone, it's not hard to imagine what it was like in those times gone by.

"Dinosaurs and the smaller creatures that lived here in John Day were real animals, not just animations," explained Park Ranger John Fiedor. "They laid eggs, raised young, fought, got sick, and died. They ate plants or, sometimes, each other!"

Fiedor said the rich fossil beds have drawn paleontologists and other researchers from around the world. Some areas of the national monument are closed to the public as research continues.

Keep your eyes open as you explore along the trails; you never know when something may catch your eye. Even if you don't come across any new fossils as you walk the trails, you will certainly see preserved animal fossils on the Island in Time Trail to Blue Basin that begins near the visitor center.

The rock walls seemed as hot as a barbecue that spring afternoon when I started on the trail. The temperature hovered around ninety degrees. Dick Creek had dried to a trickle. Along the half-mile trail I came upon three fossils under plexiglass domes—a tortoise, the skull of a saber-toothed cat, and an oreodont. With the sun reflecting off the plexiglass, it was hard to see distinctly the skull of the saber-toothed cat. But leaning closer to examine the fangs, I noticed dark blotches. Could it be the blood of an oreodont?

As our climate changes, some plants and animals are shifting northward to find new suitable habitat with food sources to which they are accustomed. Unfortunately, some may not be able to acclimate to the new habitat and may go extinct. Others will gradually adapt to survive in their new environment.

These adaptations, such as a different bill shape or altered tolerance for heat or cold, will take many years to become embedded in their genes so that offspring will retain the new characteristics. Will some new life forms eventually evolve to fill the voids left by those that could not make the transition to keep up with a warming climate?

The prehistoric animals here were replaced by bear dogs, rhinos, and camels, which were succeeded by horses, pronghorns, and peccaries, which live here today. Who knows what creatures will inhabit this landscape in 2100? I wish I could board a time machine to come back and see them.

Silent Voices

A torrent of muddy water bashes against my knees as I cautiously step over slippery stones on the bottom of raging Nine Mile Creek. My khakis are soaked and my hiking boots feel like concrete, slowing me down as I wade across the creek. Once again Utah's placid summer skies have been shattered by bolts of lightning, ripping open thunderheads on the horizon, unleashing a deluge. The skies have darkened as if the lid of a cooking pot is closing over the canyon, causing the sizzling sun to melt as it warms the rain. A flash flood has turned the creek into a river hurling rocks, trees, branches, and other debris across the packed dirt road, the only road out this end of the canyon.

Today is near the end of a week-long assignment for my husband and me to photograph rock-art panels for the Bureau of Land Management in southern Utah. We had just finished photographing the "family panel" in Cottonwood Canyon: five humanlike figures accompanied by what appeared to be a turkey and a bighorn sheep, things that were important to early inhabitants in their life in this canyon one thousand years ago.

Much of the intriguing rock art is found in Nine Mile Canyon, where it was created by people of the Fremont culture, who lived in Utah from around 800–1300 CE, when they seemed to vanish. Faced with a flash flood or drought, what would they have done? Did they have ways of coping with natural disasters that our society has lost? I ponder this question on the drive back to headquarters in Cedar City later that day in a Dodge Caravan driven by a family from Wisconsin, who happened to be on the opposite side of the swollen creek.

Archaeologists have long acknowledged that societies with less advanced forms of technology often have a keen understanding of their environment. Climatic records indicate severe, prolonged droughts from 750 to 925 CE and from 1275 to 1450 CE lasting roughly 175 years in each case, although droughts of shorter duration also occurred. Crop failures due to lack of rain along with depletion of natural resources could have caused widespread famine. When ecosystems are stretched beyond their capacity, a prolonged drought can push a society toward collapse if the infrastructure is not robust. Vivid facial expressions on some of the petroglyphs I would encounter in the next few days seemed to express these dire circumstances.

Loud, eerie moans blow across the top of a sandstone cliff soaring over Sego Canyon as I arrive alone one afternoon. The small gravel parking lot is empty. Dust swirls around my hiking boots while the sunbaked soil crumbles with each step along the short trail leading to the rock-art panels, brittle greasewood bushes snapping as I brush against them. The dry rasping voice of the wind makes me feel uneasy in this desolate place, until I notice that I am not alone.

Fading from golden peach to mauve in the glow of late afternoon, the sandstone wall forms a canvas for ghostly figures painted in reddish brown, around seven feet tall, their huge bug eyes staring at me. The trapezoidal bodies are covered with intricate designs, but they have no arms or legs. Momentarily diverting my attention, a brownish whip snake slithers in the shade at the bottom of the wall, but those huge eyes tug me back with a deep urgency. What are they trying to tell me? I glance over the line of figures, my eyes falling on a face with a grief-stricken look. Below the form is an oval drawn with a small humanlike image inside. This woman's baby has died; tears fill the corners of my eyes.

What messages do these figures have for us? Archaeologists acknowledge that the exact meaning of rock-art images cannot be interpreted today but must be examined in the context of their society at that time. Combining science, psychology, and art to study rock-art images is an emerging, controversial field. I only know what I felt gazing upon that sad face.

Archaeologists have considered these questions for ages. Reams of research into ancient cultures have been gleaned from artifacts, but is there more that we can learn by further study of rock-art images? Archaeologist Sally Cole has been studying rock art for over thirty years. "Rock art is a valuable source of information for archaeologists because it remains in place," she comments. "By analyzing rock-art styles in a certain location, we can trace patterns that may suggest interactions between cultures. It has given us clues as to what worked and what didn't work in a society at a certain place at a certain time," explains Cole. "Together with other studies, such as pollen samples, archaeologists can get a picture of what was going on at that time."

"Rock art is a remnant of another culture that is in-your-face," notes Eric Brunneman, former archaeologist at Canyonlands and Arches National Parks as well as Natural Bridges and Hovenweep National Monuments. "It drives home the point that these images were created by a person living in another time, in another culture."

According to Brunneman, if a solid date can be established through DNA testing of pigment samples, they can place the creation of the panel in a time frame and determine what was happening environmentally through pollen samples and analyzing lithic fragments. "The images were probably influenced by what was happening at that time."

Scouting the barren landscape for more rock-art images several days later, I trudge across the dusty pancake-flat hardpan toward three boulders the size of VW Beetles searching for a panel with images of a duck and what might be a fish trap or weir. Approaching the first boulder, I stop abruptly, confronted with three life-size figures etched onto the rock face. A man in the center holds the hands of two women, one on either side of him. Downturned mouth lines give the figures a sad, haunting look that I will never forget.

The Fremont survived for five hundred years because they were in tune with the land—a connection most of us have lost today. Unlike the human-induced climate change our society now faces, Fremont people didn't cause climate disruption, but they had to adapt to ongoing drought or perish.

Whether the Fremont culture died out or mixed with Ute, Paiute, and Shoshone cultures moving into their territory is still uncertain. But cultures still flourish in those harsh conditions today. Perhaps their adaptations are worth investigating as our environment steadily inches toward increasingly frequent droughts affecting our food supply, northward-shifting habitats for plants and animals, warmer and wetter winters, stronger and more frequent hurricanes along with rising sea levels.

Frequent droughts continue to affect many areas of our country. That often prompts farmers to draw from groundwater to water crops. "I think society has forgotten how much water it takes to produce food," commented Jay Famiglietté, a researcher at the National Aeronautics and Space Administration (NASA), who contributed to a study documenting changes in location of water across Earth. "We've taken its availability for granted. We're at a point in many of these aquifers where we can't take it for granted anymore. Population is too great, groundwater levels are too low."

Early cultures had far fewer mouths to feed than we do today. With more resources, most of us don't have to face starvation when the rains stop falling. More adapted to the whims of nature, early desert cultures developed systems enabling them to grow food with very little water. Perhaps there is more we can learn from those societies as our droughts increase. There may still be time for us to adapt if we act quickly and take clues from ancient cultures. But will we hear their voices before it's too late? Only if we are still. Only if we listen and let the rocks speak to us.

What Lies beneath the Ground

It was a scorching July day when we headed south thirty miles from Price, Utah, for Cleveland-Lloyd Dinosaur Quarry National Natural Landmark. The Bureau of Land Management office in Price wanted photos of this facility and others, so my photographer husband and I jumped at the chance to explore new territory while providing images for them. I had read about these enormous creatures inhabiting North America and was excited to see some authentic bones and learn more about them.

Stuck out in the middle of nowhere, across miles and miles of parched, overgrazed land, the site is recognized worldwide as one of the densest concentrations of Jurassic fossils ever found. Archaeologists have uncovered twelve thousand or more individual bones and one dinosaur egg, representing sixty different animals from seven different genera.

Vicious looking with its sharp flesh-ripping teeth, a complete *Allosaurus* skeletal reconstruction is frozen in mid stride in the visitor center, where you can look down into the quarry below to watch archaeologists excavating bones. Most are from *Allosaurous*, the largest carnivore of the Jurassic era. Researchers have also unearthed bones of plant-eating *Stegosaurus*, *Camptosaurus*, and *Camarasaurus*.

But it wasn't always dry, barren ground around here. At the beginning of the Mesozoic era, 245 million years ago, all the world's land masses formed into one large continent, Pangaea. The portion which was to become the North American deserts was then a moist, tropical lowland on the equator.

When the dinosaurs walked here 147 million years ago, there was a freshwater lake with a muddy bottom where many of them became trapped over the years. As the climate dried and the lake bottom dried up, layers of sand and mud were deposited on top of the bones and they fossilized.

Being a dinosaur freak, I could hardly wait to see some actual tracks. Though it was over a hundred degrees, that didn't dampen my spirit as we walked along the Track Tour with a ranger who pointed out footprints in the mudstone, some over a foot long. A Northwesterner used to a temperate marine climate, I was soon melting, looking around desperately for some shade. The only vegetation nearby was a scraggly bush about three feet tall where I tried to scrunch myself under its stickery, brittle branches. Ouch! No shade at all.

My conclusion that this was an utterly inhospitable environment was confirmed at lunch time, when wasps challenged me to bits of my peanut butter sandwich as I sat, briefly, at a picnic table outside the visitor center.

After a brief respite in the stuffy visitor center, we walked a short way on the Rock Wall Nature Trail, where interpretive signs describe what's been discovered and what might still be in the ground. That made me wonder about the big picture here in Utah, where natural-resource extraction has often trumped preservation of historic life, both animal and human.

Drilling for fossil fuels has probably destroyed other prehistoric animal sites as well as sacred Native American sites since one-third of the drilling for oil and natural gas is on tribal land. According to statistics comparing fossil-fuel trends in Utah, though permits to drill for oil and natural gas were down from 1,612 in 2013 to 98 through May 2017, Utah already has 5,160 producing oil wells and 7,200 producing natural gas wells.

Air pollution from coal-fired power plants is a growing concern. Of all the fossil fuels, coal is the most carbon dense and, thus, has the most impact on climate change when it is burned. Power plants in eastern Utah's Emery County, south of prime recreation areas in and around Moab, are contributing to substantial air pollution. Not only are pollen counts increasing as the climate heats up, asthma is on the rise as well.

State-wide tourism contributes $7.5 billion to the economy. Why do all these visitors come to this hot, dry place? Hiking, mountain biking, river rafting, and other forms of outdoor recreation are the draw. The magnificent red-rock scenery is a huge attraction. Mountain-bike tour operators in Moab are beginning to worry that hazy views and polluted air will keep outdoor recreationists, as well as camera-toting tourists, away.

Though the handwriting is on the rock wall, Utah legislators can't seem to read it through the smog. Short-term profits from the fossil-fuel industry are destroying much of the state's historical treasures. Wide-open spaces with plenty of sun and wind are ripe for more clean-energy installations of solar and wind power.

According to the U.S. Energy Information Administration, electricity from utility-scale wind and solar installations now provides one-quarter as much as from coal. Production from these clean-energy sources has more than tripled since 2007.

Unfortunately, the presidential administration in power in 2018 is set on a course to destroy any chance of reversing damage already done by fossil fuels and pushing us over the threshold of no return. Let Utah join with other progressive states in setting its own regulations that promote clean air and clean water in a plan for a sustainable future. This is no time to leave your head buried in the desert sand.

WINGS OF THE CONDOR

Perching tentatively on a ledge under a bridge in Marble Canyon on the Navajo Reservation, a juvenile California condor waits to hitch a ride on a warm spring thermal to glide down the canyon. Large black letters painted on a white tag on his back tell me he is H9 to condor researchers. From his perch under Navajo Bridge, H9 gazes down at the shimmering Colorado River.

Soon an adult bird, F1, swoops by, then lands on an adjacent ledge. A couple of minutes later, another juvenile and an adult appear, flying under the bridge. Spotting the two perched birds, they hesitate midair, then disappear down the canyon. Traffic whizzes by on Arizona's Highway 89 across the reservation, drivers unaware that the largest land birds in North America are soaring through the canyon below the road.

With a wingspan over nine feet, condors seem as if they belong in these canyons, carved in a parched landscape of sagebrush, juniper, and pine long ago. Wings outstretched, they glide effortlessly, following the curve of the canyon wall or the winding path of the river. The condors and the canyons have both been here for thousands of years, long before man began altering the landscape. Pieces of condor bone dating back ten thousand years have been found in caves in the Grand Canyon. That's a long time. I can't help but wonder why such a formidable flier is so rare today.

Much has changed since these prehistoric birds first graced the skies in the Pleistocene era. Originally, they ranged from Canada to Mexico and across the southern United States from present-day California to Florida. As he scanned the landscape below, the prehistoric condor saw abundant food sources in the mastodons, giant sloths, camels, and saber-toothed cats among the tropical flora below. Discovery of a carcass offered the opportunity for a feast. Condors now feed mainly on the dead elk, deer, bighorn sheep, range cattle, and horses that live in today's much drier climate.

Supremely adapted to the harsh environment, condors held their place in the ecosystem until the settling of the West, when they became victims of shooting, lead and DDT poisoning, egg collecting, and habitat destruction.

In the 1980s there were only twenty-two left in the wild. With the help of captive-breeding programs, by 2014 their numbers had risen to about four hundred, spread over various locations. But will they survive without our constant help? Challenges to their long-term survival are much greater now than in the past. Classified as an endangered species since 1969, their chief threat today is from lead poisoning, ingested while feeding on carcasses. Two tiny pieces of lead can kill a condor.

Chris Parish, condor project director for the Peregrine Fund, is heading an effort to test captured condors for lead poisoning. If traces are found, the bird is treated, then released, if possible. Another risk is outdated attitudes among ranchers who erroneously believe that condors are cattle killers, so killing them is justified.

Energy development in the form of oil, gas, and uranium extraction producing air and light pollution; road construction; as well as swelling numbers of our own species all threaten some of their habitat, even in and around national parks, such as the Grand Canyon.

Condors have more of an advantage for survival than ground-dwelling species that are reintroduced today because they can easily move to look for new habitat. For other reintroduced species, it doesn't seem to make much sense to reintroduce them if sufficient appropriate habitat no longer exists. If we, as humans, have edged them out, it doesn't seem logical to try to wedge in a creature who will still be subjected to the same environmental pressures and uninformed attitudes that diminished its numbers in the first place. Better to spend time and effort, along with any available funds, on wildlife education and improving habitat for creatures who still claim some real estate on this earth today.

With climbing temperatures hinting at the summer to come, it's quiet this afternoon, except for the occasional pickup rumbling east along Highway 89 into Navajo territory. Gliding effortlessly through the canyon, the huge birds seem like rulers of the sky. But they can only continue to claim that distinction if we allow them to reclaim a place that is rightfully theirs.

FARTHER AFIELD

AN AFTERNOON AT FORTY BELOW

Glancing over the front page of *The Seattle Times* one summer morning, I read an account from the National Aeronautics and Space Administration (NASA) that as of June 2, 2017, a massive crack in the Larsen B Ice Shelf in Antarctica had grown by seventeen miles.

Existing for ten thousand years, the ice shelf, partially collapsed in 2002, has now collapsed even more. A warmer ocean is partially to blame. According to scientists at NASA, the last remaining section of that ice shelf will probably disintegrate by the end of this decade. Thoughts race through my mind. If this happens, what will it mean for sea-level rise? What will happen to island communities or those living close to the shoreline?

Those thoughts take me back to the conversation I had with a neighbor many years ago. Howard Mason was a radio operator on the Richard Byrd expedition to that frozen continent in 1928. Byrd's expedition left California on October 10, 1928, on two ships, the *City of New York* and the *Eleanor Bolling*. The first American expedition to explore west to Discovery Inlet, then south for 140 miles across the middle of the Ross Ice Shelf, it was also the first south polar expedition to use aircraft (a Ford trimotored metal monoplane) for surveying that continent.

Surrounded by the Southern Ocean with some of the roughest waters in the world, Antarctica could be reached by ship only in November and December, usually requiring icebreaker assistance.

The ships landed at Discovery Inlet on Christmas Day 1928. "For hundreds of miles cliffs of ice and snow form a barrier about the Antarctic Continent," exclaimed Byrd upon first reaching the continent.

"Here at last was the vast, unique sheet of ice and snow that is peculiar to Antarctica."

Communication techniques were more sophisticated on this expedition than previous ones, using wireless equipment to keep in contact with the outside world, all flights, and field parties. Mason was a member of the five-man radio-communication team, which also communicated with over eight hundred amateur radio operators in the United States during the expedition. To keep in touch with the outside world, the crew received *The New York Times* over the wireless every afternoon.

Mason vividly recalled an incident on that expedition. Fierce winds rattled the field station, temperatures nearing minus forty degrees Fahrenheit. Still no word from the seven men in the geological party who had gone out hours before to investigate ice formations in a large crevasse near the camp. Mason sat at the radio receiver anxiously waiting for some word from the researchers. He hoped they would be back before nightfall, when temperatures could drop to less than minus fifty degrees Fahrenheit during extreme storms. The party shouldn't have gone out that day.

He walked over to the dining table in the center of the 24-by-36-foot building, poured a cup of coffee, then sat down at the long table. Kerosene lamps cast shadows on the faces of the group of bearded men playing poker at one end.

"I think there're coming," yelled one of the men as the door flung open and, amidst a rush of frigid air, seven frosty men stumbled in. One of them had frostbitten toes, but otherwise they appeared well in spite of ice crystals on their beards and eyebrows. Layers of clothes came off as they began to thaw out near the stove.

Soon another party would be going out on a month-long exploration to gather specimens. All the expedition parties that went out from camp took with them radio equipment so they would be in contact with base camp. Good radio communications with everyone involved was a major factor in the success of the expedition.

Mason glanced at his watch—almost time for dinner. He couldn't watch the sun go down as it had gone done a few days ago, not to be seen again for months. The main dish the previous night was penguin—yuck! Maybe this night it would be roast beef?

When he described this adventure to me, he was in his seventies. Those days he spent working in his yard, no matter what the weather. Sitting in the living room of his immaculate house with white walls and carpet, he seemed totally out of place. Here and there were touches of red, giving the living room the appearance of a giant valentine. The decor, and his life now, were determined by his seamstress wife, who would return at precisely 5:45 P.M. from her job at a department store in downtown Seattle. I used to see her once in a while returning from work, always wearing her red coat.

Still up for a challenge, Mason showed me his latest project, a pipe organ, which he was building from scratch in the basement. He had rolled the metal for the pipes and carved the wood for the keys and foot pedals. One small mistake, however, was using a vacuum-cleaner motor to produce air for the pipes. When played, the organ was less audible than the vacuum-cleaner motor!

Bringing out mementos of the expedition, he seemed proud to show me the glass photographic plates he had taken on the trip, his gold medal from President Hoover, and an autographed copy of the book *Little America*, but he couldn't find his trip journal or a number of newspaper clippings. He passed away several years after our conversation.

Looking at him then, this ordinary-looking man with eyeglasses and thinning gray hair, drove home the fact that extraordinary feats in our history were often accomplished by ordinary people. Somehow we often project the bold, the courageous among us onto a pedestal, forgetting that they were just people who had the vision and spirit of adventure to reach beyond the everyday to achieve events long remembered.

The Antarctica that the Byrd expedition investigated is much different today. Scientific teams from many countries maintain permanent research facilities on the continent. Most researchers fly in rather than taking the long journey by ship. If they do go by ship, they can reach their research stations November through February. While the Arctic is warming at three times the rate of the rest of the globe, the ice in Antarctica is melting more slowly due to the ozone hole in the atmosphere that lets in colder air, according to Cecilia Bitz, director of the University of Washington's Program on Climate Change.

Records from the European Space Agency's CryoSat-2 satellite, which records the surface shape of ice sheets concentrating on the margins, indicates ice losses in Antarctica at 160 billion tons per year for the period 2010 to 2013.

Scientists at the Virginia Institute of Marine Science predict the Ross Sea will be ice-free by 2100. At our current rate of global warming, long-term forecasts for the Antarctic do not look promising.

THE CASE OF THE DISAPPEARING TOADS

"There! Right at the edge of the tree roots, in the mud," Marvin exclaimed, slipping off his backpack and inching closer to a two-inch-long golden toad at the edge of a tiny pond. As I moved closer, the tiny guy hopped into the underbrush.

On a visit to the Monteverde Cloud Forest Reserve on a trip to Costa Rica many years ago, my husband and I struck it rich by having Marvin Rockwell as a tour guide, leading us on a hike through the cloud forest as if it was his backyard. Thrilled to see so many tropical plants and birds on a brief walk by ourselves through the cloud forest, we were curious about the golden toads. Were they really golden colored? Why were they so hard to find? If anyone could find a golden toad, it was Marvin. It seemed a perfect environment; I wondered why they were so rare.

"When I got here, there were hundreds of these golden toads," he recalled. "Hardly see them anymore." Tall and lanky with a few tufts of gray hair, Rockwell was one of the original settlers of this region of Costa Rica, seeking a peaceful place to live outside of the United States. A handful of Mormons drove from Alabama to Costa Rica in 1959, settling in Monteverde. The area was perfect for a dairy farm, one that is still producing today. The delicious caramel sold in the small retail shop on the farm was divine!

The 35,089-acre reserve protects what remains of the cloud forests that once covered the Cordillera de Tilarán in northwestern Costa Rica, located about 110 miles northwest of the capital city of San José. According to the tourism website, this misty greenbelt provides protection for 2,500 plant species and 400 bird species, including 91 species that nest in North America and migrate to or through the cloud forest. Dazzling red and green resplendent quetzals and other colorful birds stood out among the greenery. Of the 100 mammal species, including jaguars, ocelots, peccaries, and agoutis, the only ones we saw were peccaries and agoutis. Of the 1,200 species of amphibians and reptiles, 71 of them are snakes. Voracious mosquitos are among the tens of thousands of recorded insects.

We were surrounded by exotic plants that climbed up trees and each other, as if trying to engulf us all in their verdant, leafy world. Flowering bromeliads pushed the bottoms of their stiff leaves together to catch moisture, forming tiny pools—perfect for the vast array of small frogs living in the cloud forest. Along the muddy trail, I watched platoons of determined army ants, carrying bits of leaves aloft like banners as they marched over anything in their path.

Now considered extinct, the golden toad was last seen in 1987, a couple of years after our visit. Males were a brilliant gold, while females were an earthen color with dark spots. These insect-eating amphibians spent half their time on land and half in water, where they bred.

Scientists believe the demise of the golden toad may have been due to chytridiomycosis, an infectious disease caused by the fungus *Batrachochytrium dendrobatidis*, which affects amphibians. These elusive creatures were good indicators of environmental health since they lived in both terrestrial and aquatic environments.

The cloud forests are normally warm and humid. But an El Niño event between 1986 and '87 caused the climate to become warm and dry, a perfect environment for the fungus to flourish.

Habitat loss, climate change, as well as expanding and new infectious diseases are some of the main challenges amphibians are facing today. A warming environment may promote an increase in the fungus or other diseases that can cause too much water and oxygen, along with pollutants, to be absorbed through their skin.

As our climate warms, which species will adapt and which will disappear? Which invaders will push into the new habitats created, edging out the few natives left? According to researchers who study the climate today, a warming climate fosters more bacteria and diseases that are harmful, often deadly, to plants, animals, and humans. As Drew Harvell, professor of ecology and evolutionary biology at Cornell University puts it: "A warming world is a sicker world."

FAIREST ISLE OF THEM ALL

The soft wool sweater-vest felt as if it was caressing my hands. Hidden under a humble cotton cardigan, the vest felt warm even though I was holding it, not wearing it. Earth tones used in the horizontal rows of geometric designs reminded me of the stark beauty of the windswept island where it was created: the cream color of the sea campion wildflowers clinging to the cliffs; the tawny color of the soil; the dark brown of the rocky cliffs where seabirds nest.

Holding my Fair Isle vest now summons a flood of memories of our visit long ago. My late husband, Jim, and I were planning a trip to England, Ireland, and Scotland, so Fair Isle, off the northeast coast of Scotland, would be sort of in the neighborhood. On the map it looked so close to the mainland, how could we not go? The lure of its remoteness, abundance of seabirds, and legendary Fair Isle sweaters was too much to resist.

What was it like to live in such an isolated environment, away from hordes of people and traffic? Now was my chance to find out.

After a couple of hair-raising weeks with Jim driving on the right-hand side of the road, broken up every few days with relaxing, scenic train travel, we boarded a Loganair flight in Edinburgh for the two-hour flight to Lerwick on Shetland Island. The following day after a half-hour flight from Lerwick, we landed on Fair Isle.

North of the Scottish mainland, this three-square-mile island is ringed with rocky outcroppings forming cliffs, perfect for nesting seabirds. Ancient sandstone forms sea stacks, arches, and cliffs around the flat-topped island, which is riddled with peat moors. Both grey and common seals are at home in the rough waters of the North Atlantic surrounding the island.

Weather was mostly calm and cloudy the week we were there in August, although the sun peaked out a day or two. Whistling wind, cries of birds, and the foghorn were the only sounds most of the time. Sheep from the many island flocks joined in the chorus on occasion. There were few trees on the windswept island, but wildflowers reminded us of the season, despite the cool temperatures. Clumps of tiny pale mauve blossoms hid from the wind between boulders while clusters of pink and white angelica and yellow bog asphodel stood out on the hilly moorland in the north of the island.

Our room at the Fair Isle Bird Observatory (FIBO) guesthouse was sparsely furnished—just two twin beds, a desk, and one chair. Though it was summer, there were only a few guests. The dining room served simple fare for breakfast and dinner. Lunch was on your own. Toward the end of the week, it was hard to face another meal of meat, boiled potatoes, and well-done cabbage, carrots, and turnips. Salad was a treat since the growing season is so short. Warm, tasty breakfast scones were a highlight each morning.

A new FIBO guesthouse, built in 2010, welcomes more visitors, along with a steady stream of researchers. There is no television, but there is free Wi-Fi. It was the only place to stay when we were there; now there are other accommodations.

We rented a beat-up car for two days to get an overview of this remote bit of Scotland. Most days we just walked. Daily hikes took us to all parts of the island: the cliffs to photograph seabirds, into town to chat with the local folk, and to a community gathering at the far end of the island one evening. Late into the night the small bungalow was jumping with fiddle tunes for the Shetland folk dances. We left around 11:00 P.M., knowing we had another two-and-a-half-mile walk, in the dark this time, on a narrow rocky path with our flashlight guiding the way back to the guesthouse.

Many islanders wore Fair Isle sweaters, called *jerseys*. The distinctive style of knitting began in the 1600s, when islanders made them to barter with passing ships. Flocks of Shetland sheep provide the wool used to make yarn woven into sweaters with traditional Fair Isle patterns. Though many clothing manufacturers now produce replicas, the only authentic ones are knit on Fair Isle. Sweaters are custom-made and hand knit by only a couple of knitters on the island; they are not cheap. Lower-priced ones using the traditional patterns are knit by machine on mainland Shetland.

My Fair Isle vest was hand knit by Florrie Stout, whom I first met when she was giving a knitting demonstration at the local community center. Next to her stood a table where she had arranged small knit samples of sweater patterns.

"Ere," she tells a visitor in her clipped brogue, "watch 'ow I graft the neckband onto the bod o' this jersey." Wool from Shetland sheep comes in shades of gray, white, black, tan, and a reddish-brown. The wool is sheared in June, then sent to Lerwick to be spun and dyed, although many customers prefer the original color of the wool to be used to make their garments.

Stout measured me for a vest and Jim for a cardigan. About six months later, we were delighted to receive them in the mail. Today the name of a person ordering one of the distinctive sweaters will go on a waiting list that is three years long!

Keeping up with daily chores around the farm creates a quiet lifestyle. One popular annual event is the visit of a dentist and his assistant from Lerwick, bringing all his equipment, including a dental chair.

Luckily, we didn't have to make time to visit the dentist during our visit, which left more time for birding. Almost 390 bird species have been recorded on the island. Loaded down with camera gear, we headed out each morning eager to find some of those birds. But we didn't realize a few of the aerial acrobats would come to meet, or warn, us, not long after we headed out from the guesthouse each morning.

"Wow, that was close!" I yelled to Jim as an arctic skua swooped out of nowhere, dive-bombing me within a couple feet of my head—shrieking a stern warning to steer clear of its nest on the ground nearby. The nest of this blackish summer visitor was easy to miss; just a depression in the dirt of the flat landscape with its cover of green stubble.

Around the size of a large gull, arctic skuas are particularly adept at stealing food from terns and gulls. Why hunt for your own food when you can steal a meal from a neighbor?

After that first encounter, we walked more warily, ready to hoist our tripods over our heads to discourage an attack by these extremely aggressive birds. They certainly kept the island's many rabbits on the run.

Hiking to the cliffs where the seabirds nest was an adventure each day. We never knew what would show up. Though nesting season was over, we still saw lots of birds, including gull-like creamy-white fulmars, kittiwakes, graceful gannets, razorbills, both horned and Atlantic puffins, and more.

Looking over the cliff edge to the rocky shoreline below, we watched black oystercatchers, ringed plovers, and dunlin searching for a tiny morsel at the edge of the surf. Offshore, shearwaters and storm petrels patrolled the waters.

By the time we left, the third week of August, it was getting dark earlier and the temperature was dropping. Gusts were getting stronger. Only the fulmars were left on the cliff, some with juveniles in their gray down coats. A few puffins stood around their burrows, late leaving due to a food shortage. Shags (similar to a cormorant), fulmars, and puffins eat mainly sand eels, but fishermen also go after them. An essential food for the birds, the small thin eels are made into fish meal for animal feed and also used in fertilizer.

Food shortages due to overfishing and a warming ocean, along with variation in the usual arrival dates of spring and fall weather, are having a noticeable effect on migrating birds. Since 1948 the Fair Isle Bird Observatory has kept records of the migrating and nesting seabirds and land birds.

The effects of climate change on the birds have been apparent for several years. Swallows, one of the species of land birds recorded, are now arriving three weeks earlier than they did in the 1950s and leaving earlier.

According to Will Miles, PhD, of the Fair Isle Bird Migration Project, this could be due to "changing climate and weather patterns, change in the summer breeding range, or other factors."

Bird researchers calculated that the breeding populations of fulmars have plummeted from 43,000 pairs in 1996 to 32,061 in 2016. Some deaths were caused by storms or food shortages, but others were due to the seabirds ingesting bits of plastic floating in the waters. A number of the dead birds had up to thirty small pieces of plastic in their stomachs.

Sitting in the stormy North Atlantic, Fair Isle has taken advantage of wind as a source of energy since the 1980s. The two aging wind turbines installed during those years, along with diesel generators, have not produced enough electricity to keep the lights on twenty-four hours a day. That has meant no electrical power from 11:00 P.M. until 7:00 A.M. each day when there was not enough wind.

Fortunately, funding came through in July 2017 for three new, larger wind turbines along with a solar array and battery storage. Hopefully, a more consistent, reliable power source will attract more people, expanding the population beyond the present fifty-five, and allow businesses to grow the infrastructure of this already-green island community.

As our Loganair flight took off from the tiny Fair Isle airport on the day of departure, I watched the island disappear into the clouds. How would all those millions of nesting seabirds survive if their food supply continued dwindling in a warming ocean? Will more adventurous residents heed the call of island life, willing to put down roots in such a remote outpost?

This is prime time to install the new power sources that will generate more reliable electricity to homes and businesses. More communities should follow suit. It seems that nature is in the driver's seat now, and we'd better ride along or suffer the consequences of not taking action in the face of an oncoming disaster.

FASCINATED WITH THE FLOW

Make a date with Madame Pele, legendary Hawaiian goddess of fire, and she may show you the time of your life. Fiery rivers of red-hot lava cascade down the mountainside, incinerating trees, homes, or anything else in their paths. Pillows of molten lava creep a few yards from your feet, inching closer to the cliffs where they will tumble, boiling and sputtering, into the sea. A hot wind engulfs you in the fumes of her sultry, sulfurous breath.

But Madam Pele is unpredictable. She may present you with an unforgettable spectacle or she may stand you up, leaving you alone amidst miles of cold, black lava hardening in swirling patterns, attesting to the fury of her last encounter.

One of the most thrilling places in the world to experience this awe-inspiring phenomenon is at Hawai'i Volcanoes National Park on the island of Hawai'i. Kīlauea volcano has been erupting off and on since the 1980s. But why are people rushing toward this erupting volcano instead of fleeing in terror? We were lucky enough to find out.

When we applied for volunteer writer and photographer positions at the park, my late husband, Jim, and I had no idea we'd be drafted into eruption-flow duty when the park service came up short-staffed one day. It turned out to be terribly exciting.

We had just spent several days gathering information and photos for a trail guide. Jim, fearless photographer that he was, captured many close-up, spectacular lava images without getting his shoes melted—an amazing feat!

But now, given temporary "official" status for the day, we donned yellow hard hats and orange safety vests, tucking a short-wave radio in a vest pocket, ready to keep visitors back from dangerous places we had been in the days before.

After just a few minutes at my post, I spotted a visitor clambering over the sharp lava rocks beyond the barricade, getting dangerously close to the flow. Clutching the *Fact Sheet for Eruption Flow Duty* in my hand, I asked a woman with a camera to step back behind the barricade. She was obviously determined to get a dramatic shot of the fiery red lava flowing toward the sea. Defying my request, she continued gingerly stepping over the jagged rocks toward the river of lava.

Hot, humid air laced with sulfur fumes can tamper with a person's common sense when viewing flowing lava this close. While it seems silly to risk injury or even death for the sake of a few photos, I don't really blame her for trying to bend the safety rules as posted. This is a once-in-a-lifetime experience for most people.

"We may get one thousand to fifteen hundred visitors or even two thousand visitors a day when there is a lot of flowing, red lava," said Mardy Lund, former interpretative ranger. "They come from all over the world, and many come expecting to see flowing lava. Some people get very angry when there isn't any volcanic activity while they are here."

Hawaiian volcanoes Kīlauea and Mauna Loa are the largest shield volcanoes on Earth, built by successive layers of lava. Kīlauea rises more than twenty thousand feet above the ocean floor, with 4,190 feet above sea level. Scientists estimate that Kīlauea may have emerged from the sea fifty thousand to one hundred thousand years ago. As it reached sea level, erosion began taking place and the summit collapsed, forming a caldera.

The smooth shield profile was altered as a cap of lava built up on top, producing steeper slopes and a jagged summit. The caldera at the summit is two miles wide by two and a half miles long. Kīlauea's main vent is located below a collapsed crater, Halemaʻumaʻu, at the southern edge of the summit caldera. White-tailed tropic birds are often spotted circling the rim of Halemaʻumaʻu, seemingly immune to the noxious fumes.

The viscosity or liquidity of the lava, determined by its chemical composition, temperature, and the amount of gas present, influences the type of eruption likely to occur: either explosive, generating steam, ash, and other pyroclastic material, or lava flows. Highly viscous lavas contain more gas, producing destructive, explosive eruptions such as the one in 1980 on Mount St. Helens in Washington state. Lava from Hawaiian volcanoes is less viscous, producing more of a slow flow, like syrup.

In fact, Hawaiian volcanoes are gentle giants. These lava flows, though they may be one to two million cubic yards of molten rock flowing per hour at temperatures up to two thousand degrees Fahrenheit, are predictable.*

The eruption I was watching began in 1983 on the east rift zone. For three and a half years, there were forty-seven brief episodes of lava spraying upward like a fountain, thrown 1,500 feet in the air. Since the lava had been erupting for several years, there was no pressure buildup, but rather a constant flow from a lava lake through a tube seven miles long to the ocean.

Ongoing eruptions of Kīlauea have provided scientists the opportunity to document the formation of pillow lava from submarine lava flows that enter the Pacific through a well-developed tube system. Most of the lava forms billows of pillow lava on the ocean floor.

Around two million visitors a year are attracted to the erupting volcano. While all those visitors pump dollars into the local economy, some put themselves at risk. Flows are accompanied by fumes of hydrochloric acid, especially harmful to people with respiratory problems. Visitors

---

*On May 3, 2018, Kīlauea began the largest eruption in a century. Fissures hundreds of yards long opened up, spewing toxic gases, steam, and lava into the air. Several residential areas were evacuated. As of this writing, volcanologists were expecting an explosive eruption from Halemaʻumaʻu crater, with a 20,000-foot-high ash cloud possible.

going beyond warning barriers may get burned by particles of lava hurling through the air or by walking over thinly crusted areas.

Wearing an official outfit, adopting an official attitude, I couldn't believe that some visitors were exploring the jagged volcanic landscape wearing flip-flops, sandals, or other flimsy footwear, which can often result in sprains or broken bones from falls. Stepping or falling on sharp volcanic glass can also result in nasty cuts.

Even tennis shoes are sturdier, though rubber soles can melt if visitors walk on top of recently dried lava. Barriers placed at a safe distance from an active flow must often be moved several times a day as the flow changes direction. But time and again, some visitors ignore barriers in their attempts to get spectacular images—they are so fascinated with the flow. I got more than a few angry looks that day trying to keep people behind barriers.

Stepping carefully over jagged lava dried ages ago, I was amazed to see lichen along with green sprouts of native ohia and 'ama'u fern coming up in cracks in the barren, rocky landscape. Other pioneering plants will follow suit. Wind will carry soil and seeds into crevices of the hardened lava. Droppings from birds and small animals will deposit seeds into the soil that will soon sprout in the humid atmosphere. Over time, native shrubs and trees will take root.

Scientific studies of the ongoing eruption have aided in determining safe areas where homes may be built. However, lava flows don't always follow their predicted routes, sometimes leaving a path of destruction in their wake.

The natural world can slowly recover over time from a catastrophic event such as a volcanic eruption. But can the earth adapt to an explosive population using more and more natural resources like there is no tomorrow? Can she adapt to rapidly increasing levels of carbon dioxide filling the atmosphere and the oceans, disrupting life as we know it? If so, Earth will be a very different place.

FRANCE'S ANCIENT ARTISTS

September 11, 2001, while millions of horrified people across the country watched a scene of incredible devastation unfolding, I was underground watching scenes of incredibly elegant artwork produced nearly twenty thousand years ago. Oblivious to the twin towers of the World Trade Center crumbling, I stood in awe, viewing magnificent prehistoric creatures painted thousands of years ago on cave walls in southern France. Who were these people? What did these images represent to them? Did the images shed some light on how early civilizations survived during periods of climate change in their time?

After seeing the images in books, I was determined to see the originals before these fragile treasures were closed to the public. Images in world-famous Lascaux could only be viewed in Lascaux II since the original cave was closed to the public. Some of the others were still open, but for how long?

The Vézère Valley in the Dordogne region of south-central France claims one of the world's highest concentrations of prehistoric cave art. Rivers, lush forests, rocky shelters, and caves hidden in the limestone cliffs made it an ideal location for the Cro-Magnon people. The Vézère River running through the valley would have been a magnet for animals from the region, making them prime targets for hunting.

Many of the cave-art sites are clustered around the tiny town of Les Eyzies. Unfortunately, the world-renowned prehistoric museum was closed for renovation when I was there. As it turned out, the experiences I had visiting three caves, Lascaux II, Grotte de Rouffignac, and

Grotte de Font-de-Gaume, were more powerful than any I would have had viewing artifacts in the museum.

The warm autumn day outside turned chilly and damp as our open-air tram entered into Grotte de Rouffignac. The tour guide was silent for a few moments as eight tourists craned their necks to take in the huge images on the ceiling of the cave. The enormous mammoth painted above me looked so realistic I could almost feel his steamy breath wrapping me in a smothering blanket, his two long curved horns skewering me into a human shish kebab. If this guy bolted for the cave opening, I'd be history! Carefully drawn on the ceiling by an ancient artist nearly seventeen thousand years ago, the shaggy beast still radiated energy through the layers of cool limestone that sent shivers up my spine.

These stunning, enigmatic images, executed in a sophisticated style often incorporating natural aspects of the rock wall, were created by ancient hunter-gatherers who flourished thirty thousand to fifteen thousand years ago. They used their artistic skills to express their thoughts and feelings in the form of mythological beasts, animals for the hunt, images of their gods, and graphic drawings of events in their world.

Every stroke, every line, had meaning for the artist. Scholars studying cave paintings recognize that drawing definitive conclusions about motivation for creating the images is impossible today. However, they agree that the drawings were an expression of psychological and social needs: a need to warn others of danger; a need to invoke blessings from a hunting spirit for the power to kill the mammoths, wooly rhinoceros, deer, horses, and other animals they encountered. For them, killing animals was a matter of survival, although some of the animals were hunted nearly to extinction—a situation often perpetuated by trophy hunters and poachers today.

These Cro-Magnon artists excelled at realistically depicting the animals they hunted and those they feared. Walls and ceilings of Rouffignac are covered with 154 images of mammoths along with ibex and horses. It was so exciting to see actual claw marks from cave bears along the bottom of the cave walls. An extinct species, they denned in the caves in winter. At some of the other cave-art sites are images of short, stocky horses, wooly mammoths, cave bears, ibex, bison, red deer, and a few humans, ranging in size from about eight inches to over ten feet.

Natural elements from the earth mixed with animal grease provided materials for the paint used by the cave artists. Black could be made from manganese dioxide; yellow, from iron oxides; and red, from ochre. With no charcoal pencils available, the Paleolithic artists probably used their fingers for drawing. Paint could have been dabbed on the wall with tufts of hair or moss. A hollow piece of bone made a dandy blowpipe to spatter paint directly on the cave wall. A sharpened stick or bone became an engraving tool.

A local tour guide enlightened me on different cultures who had used the caves. "These cliffs and caves have provided refuge for different cultures throughout the ages," said Christine Bonhomme. She pointed out the lines of small square holes at the base of some of the cliffs, explaining that they were used to prop up timbers supporting the roofs of medieval homes built at the base of the rock wall. "The caves were also used throughout the religious wars, in the Middle Ages, World War I, and by members of the French Resistance in World War II," she added.

Art this old can take your breath away with its elegant, sophisticated lines and finely crafted details. Like most fine art, it's extremely fragile. Hidden in caves for centuries, the images have had some protection from wind, rain, and human disturbance. Climate changes over the centuries have caused fluctuations in temperature, moisture, and humidity. Bacteria that entered the caves on the clothes and shoes of visitors has damaged some of the original paintings. When environmental conditions are altered or foreign organisms introduced, the art begins to deteriorate.

France's cultural-heritage officials perform a balancing act between allowing some public access to the caves while still preserving this precious heritage for scientific research. Lascaux, one of the best-known painted caves, was closed to the public in 1963, when bacteria and carbon dioxide from visitors' breath began to destroy the paintings. Now visitors can tour Lascaux II, a detailed replica of the original cave, which took ten years to complete. It certainly looked authentic to me. Both Grotte de Rouffignac and Grotte de Font-de-Gaume that I visited are still open to the public, but allow just a few visitors each day.

Migrating from Africa at the end of the last ice age, these early people lived through climate fluctuation lasting around one thousand years when temperatures gradually became warmer and more humid. They survived these changes because they knew their environment intimately and adapted to extremes of hot and cold. They were familiar with the life cycles of the plants and animals around them.

Living amidst great biodiversity, Cro-Magnon people ate a wide variety of plants and animals. According to Marilyn Walls, who holds a master of science in nutrition, today we get our food mainly from five animals and twelve plants. Loss of biodiversity will greatly affect our food supply. As biodiversity dwindles due to a changing climate, habitat loss, and other causes, how will we adapt to a shrinking food supply?

Today we've lost a lot of the adaptability that helped early societies survive. We try to control nature rather than adapting to it; even experimenting with schemes to control the weather. But if global temperatures and sea levels rise, how will we cope?

History provides many examples of larger, more efficient and cohesive societies surviving major challenges while smaller, divided ones suffered. In his book, *The Cave Painters*, Gregory Curtis talks about one theory as to why Neanderthals disappeared around thirty thousand years ago as the ice age became more severe, while Cro-Magnon people survived. They were a larger group and more organized. Learning from others and helping each other made them a stronger society.

What kind of legacy will we leave for future generations—one with clean air and water? Will the future landscape include lush forests, free-flowing rivers, and abundant wildlife? We'll leave painted art and photographs showing our world as it is now, but how much of the natural beauty we enjoy today will still exist for our great-grandchildren's children? It's up to us to protect it today for ourselves and for those to come.

Haida Gwaii

Salt spray pelted my face as I gripped the rope handle around the sides of our Zodiac while it slammed against swells on the heaving sea. Earlier that day we had boarded a ferry in Prince Rupert on the west coast of British Columbia, Canada, for the comfortable ride across Hecate Strait to Haida Gwaii, just over sixty miles west of the mainland.

We docked at Skidegate on Graham Island, where we climbed aboard a seaplane that took us to Kunghit Island, the site of our base camp for a week of exploring. The adventure began as eight urban-dwelling adventurers gingerly transferred our gear and ourselves from the seaplane into an inflatable boat as it bobbed up and down on the choppy sea for the ride to shore.

Chilled to the bone and dripping wet, we reached our eco-adventure base camp in about twenty minutes, but it seemed twice as long. Nestled in the mossy rain forest, our compound included six large sleeping tents, a dining tent, a drying tent for our constantly wet clothes, and a latrine tent, all tucked under towering red cedar, Sitka spruce, and western hemlock. A thick blanket of moss covered every inch of ground. Though it was August, chilly mist and fog most days made it feel more like late October back home in Seattle.

Each evening at dusk, a Sitka black-tailed deer visited camp, munching on moss as it casually glanced at the colorful, animated intruders. Bald eagles flew overhead while a Peale's peregrine falcon screeched at us from its perch in a hemlock. Life teemed around us. Each day, in our inflatable boat, we motored around some of the many small islands, clambering on shore at times for a closer look.

Tufted puffins, with outrageous orange bills and yellow hair tufts, peered at us over boulders. Curious harbor seals circled our boat. Sea lions slid off their rocks as we approached. Flotillas of pigeon guillemots, black seabirds with bright red feet, passed by our watercraft. With the motor shut off, we quietly drifted into Rose Harbour for a thrilling view of three stunning orcas feeding.

The second day of our visit, we headed to Anthony Island. Approaching the island, we caught glimpses of some of the ancient totem poles hidden in the forest. The frogs, beavers, ravens, and other figures carved into the weathered gray totem poles seemed to be watching us....

Our group of eight intrepid explorers was joined by four tourists swooshed in by chartered helicopter. We all stood quietly while Wanagun, archaeology warden on Anthony Island, told us about the ancient village of Ninstints and the Haida people who lived there.

"Please be careful where you walk," he advised as we gazed at the ancient totem poles around us. "The old house frames crumbled. Don't step on anything wood. You could be stepping on artifacts. Above all, please respect this site that is sacred to the Haida people."

The Haida have inhabited the Queen Charlotte Islands, now called Haida Gwaii (Islands of the People), for about twelve thousand years. There were two dozen villages around Moresby, the south island. Not many remnants of those villages remain except at Ninstints, Skedans, Tanu, and a few other sites. Contact with European fur traders brought epidemics of smallpox and other diseases that wiped out most of the population by 1860.

Exposed to the elements, the fragile totems and other wooden artifacts are deteriorating. Wanagun spoke about Ninstints, a World Heritage site, one of three abandoned villages on Anthony Island dating from the early 1800s. Towering cedar and spruce trees swayed in the wind. At the edge of the bay, sixteen gray, weathered totem poles—keepers of family histories and legends—leaned, staring out to sea. "These totems may last only another twenty years," he commented.

Standing between two of the worn totems, with the wind howling around me, I faced the sea and closed my eyes. I could almost hear Haida songs and drum beats from some long-ago potlatch.

A couple of years after our visit some thirty years ago, the Haida Nation officially gave back to the Canadian government the name Queen Charlotte Islands and became Haida Gwaii, in a statement of their sovereignty. Not long after the name change, south Moresby Island became Gwaii Haanas, a new national park and national marine reserve.

When we visited, the destination was off the beaten path and received few visitors. Now there are flights from Vancouver or Prince Rupert to Skidegate, where visitors can find a selection of motels, guest houses, and B&Bs, as well as rental cars to explore Graham Island or sign up for a tour. It's best to plan ahead and reserve a tour to Gwaii Haanas National Park, since it is only accessible by boat or plane.

Climate change holds the threat of damage to some of the historic sites. According to a report done by researchers at the University of Victoria, B.C., the north end of Graham Island is most at risk for sea-level rise. The east side of the island is also at risk, with the west side of the island at moderate risk.

Sea-level rise and extreme weather events will probably damage some low-lying communities. The coastal highway on Graham Island will be flooded on occasion. Air and sea routes used for delivering goods from the mainland could be affected even more than they are now by the increase in severe weather events.

Dr. Robert Bindschadler, a National Aeronautics and Space Administration scientist emeritus, commented in a climate-change talk I attended that he is more concerned about extreme weather events than sea-level rise. His rationale is that we can plan for sea-level rise, but the increase in devastating storms, hurricanes, extremely heavy rainfall, and other weather disasters is often sudden and completely out of our control.

To address the future flooding of roads and communities due to sea-level rise, the Haida are continuing to work on plans that will allow the communities to function and obtain necessities under extreme weather conditions.

The marine environment as well as rivers and streams have been a source of food for centuries. Warming waters, along with other factors, are having an impact on this food supply. Herring, a staple for the community, has been in decline since 1995, partially due to changes in their marine environment.

The once-abundant Pacific salmon have been decreasing in number since the early 1990s due to overfishing, habitat destruction, and warming stream temperatures. In partnership with Parks Canada, the Haida are addressing this issue. A project in the national park aims to restore salmon streams damaged by poor logging practices in the past that made these waterways more susceptible to warming temperatures. Haida youth are participating in the project, while being trained to conserve and protect the lands and waters of their heritage.

The Haida are a proud, strong, resilient people, accustomed to power outages, interruption of ferry and flight service, short-term food shortages, and occasional flooding of coastal roads. A close-knit community, they are keenly aware of their environment and can rely on traditional abilities such as food gathering and stockpiling, hunting, and back-country skills when needed.

They continue to stand up to the Canadian government for control of logging in their homeland. Haida solidarity against logging on south Moresby Island played a significant role in the establishment of the national park.

Living in the unpredictable Pacific, their emergency kits are always ready. The Haida Nation has overcome many obstacles to preserving their land and heritage in the past. The onslaught of climate change is bringing even more challenges.

Risky Business

As the LIAT airlines plane began its descent into Hewanorra International Airport on the tip of St. Lucia, in the Caribbean, my nose was pressed against the window. Mesmerized by the lush green island floating on an aqua sea, it looked like paradise. A freelance writer for many years, I relished the discoveries each new article presented. For this one, I had been assigned an on-the-ground report on cacao production at the plantation owned by a large chocolate manufacturer in the United States. My husband, Jim, a professional photographer, was going to take photos for this article.

Poverty and politics in cacao-growing regions worldwide contribute to an unreliable supply of cacao beans. Incidents of forced child labor on some plantations, particularly on the Ivory Coast, have plagued the industry. Progress is being made in the industry to abolish this shameful practice. Hopefully, this experimental plantation on St. Lucia will act as a catalyst to encourage stable cacao production while also improving the lives of the plantation workers.

Their primary mission was to create insect- and disease-resistant strains of the beans, train local farmers to grow them, and pay the growers a fair wage, helping to stabilize the supply while making strides to improve the lives of the workers. I was especially interested in getting an in-depth account of how this would work, from the bean to the bar, so to speak.

Soon after the plane touched down on the tarmac, we stepped out into the sweltering sauna of St. Lucia in July. My contact for this assignment was Philip Rousseau,* a tropical agronomist. We were met by a tall, gray-haired man, dressed in white slacks, embroidered light-blue cotton guayabera, and a Panama hat, who motioned us toward his white van. Philip chatted about the natural history of the island as the van sped along a narrow winding road past small villages surrounded by banana and spice plantations.

A spice garden full of cinnamon, vanilla, and allspice plants spread in front of the graceful white two-hundred-year-old sugar-plantation estate, a porch sprawling across the front. When the sugar industry died out, the plantation was converted to 238 acres of cacao trees. Cacao flourishes in this climate, but I was wilting fast. Our room was stifling. After unpacking a few clothes, I opened the shutters, hoping for a breeze. Then we headed to the dining room where Maria,* the cook and housekeeper, had prepared a typical island meal of chicken cooked with manioc and papaya. After dinner Philip, Jim, and I walked to the old sugar mill, not far from the house. Rusty now, it was in its heyday when sugar was king in the 1700s.

After a long day, we finally retired to our room to pack gear for the next day's visit to the plantation. But, as soon as I turned on the small bedside lamp, a cloud of cacao moths filled the room and began landing on me! Twitching about, I brushed the creepy insects from my head, my arms, my clothes. Jim didn't seem to be bothered by them. Switching off the light, I washed and brushed my teeth quickly in the dark and jumped into bed, pulling the sheet over my head. A couple of hours later, the air seemed still, so I pushed the sheet down from my head and dozed off.

Over the next couple of days, I learned what goes into the birth of my next silky-smooth dark-chocolate bar. Creole-speaking black plantation workers cut the yellow and orange cacao pods from the trees and loaded them into baskets, which the women carried on their heads to

---

*Name has been changed.

an ATV. The pods were poured into boxes on the back of the ATV that would take them to a truck where they were loaded up for the ride to the processing sheds.

There, the pods were whacked open by machete and the slimy white beans scooped out onto racks and covered with banana leaves, where the beans began to ferment, developing the characteristic chocolate flavor. Next, they were spread on large wooden trays to dry in the sun for a week. While drying, the beans were continuously turned. Dried, cleaned, and bagged, the cacao beans were trucked to the port city of Castries for export.

Armed with notebook and pen, I shuffled through the fallen cacao leaves, hastily jotting notes as Philip rattled off statistics about the plantation. Clad in colorful cotton dresses or pants and shirts, the workers seemed shy. I wondered if they thought I was reporting on how efficiently they were working. One hefty woman with a basket of cacao pods on her head cracked a slight smile as she sauntered past me. Like some of the others, she wore rubber boots for protection against snakes. Many workers were barefoot.

With the cacao trees shaded by taller trees, this plantation seemed well planned. According to the Rainforest Alliance, the lower story of an evergreen rain forest is the natural habitat for cacao trees. The trees are not only shade tolerant, but the beans are of better quality than ones from trees grown in full sun. But disease, insect infestation, such as the cacao moths, and hurricanes are a constant threat. In 1980 a hurricane wiped out 10 percent of these cacao trees and damaged 50 percent more. Hurricane Irma, the monster hurricane in September 2017, passed well north of St. Lucia, so the island was spared, this time.

Philip's work, along with that of other agricultural researchers, couldn't be more urgent. With a warming climate, known diseases and new ones for plants, animals, and people, are predicted to spread and worsen. Hurricanes are already increasing in frequency and force.

According to a report from the Intergovernmental Panel on Climate Change, increasing drought, occasional excessive rainfall, and other extreme weather events may affect the quantity as well as the quality of cacao beans.

The afternoon before our flight home the next day, I sat in a wicker chair on the porch, tapping my notes into my laptop. The wicker was pressing a pattern into my back through my soggy shirt. Ripe mangoes thudded to the ground from a tree by the porch while a tiny hummingbird, resplendent in its iridescent blue, green, and violet outfit, flitted among red hibiscus. Swigs of tangy limeade and nibbles of Maria's coconut cake helped keep my squishy brain focused on my notes. Though I don't think I could ever get accustomed to this tropical climate, through research, this plantation may be able to adapt to changing growing conditions.

Clouds began to fill the azure sky as the plane lifted off. Let's hope cacao plantations can cope with a changing climate, or the chocoholics among us are in real trouble.

TIGHT GENES

"Carl W. Johnson, 1899, Västra Torsås, Sweden." The name jumped out at me from the silver plaque mounted on the base of the statue of Leif Erikson in the plaza named for this Norwegian explorer at Shilshole Marina in Seattle. That's the year my grandfather left his birthplace, never to return. When I filled out the paperwork to have his name added to the plaque, I knew he would want to have the American spelling of his name, rather than the Swedish: Karl Wilhelm Johannason.

It was a cloudy fifty-degree weekday on the waterfront. The salt air, the bay, the boats, all these things made me feel closer to him. He was just eleven years old, coming to America with an uncle on a steamship out of Gothenburg, landing at Ellis Island. I could almost see the ship coming into this marina that day, an eager young boy peering over the railing.

Between 1840 and 1893, 1.3 million Swedes came to America, mainly for economic reasons. What were his first impressions of his new homeland? Exciting? Overwhelming? Scary? I often wondered.

Once in America, Grandpa's uncle took off soon after depositing him with a Swedish farm family in Iowa. From farmhand to nursing-home worker in Iowa and, later, to Bremerton, Washington, where he worked in the shipyards during World War II, to boilermaker for the Alaska Railroad; his life took many turns.

After about twenty years in Alaska, he and Grandma drove his old Packard (picture a baked potato on wheels) to Vancouver, Washington, to be closer to our family. I harbor fond memories of him working in his vegetable garden behind their neat white house, wind-blown white hair atop his sun-tanned face. He loved fussing around in that garden; his carrots were the sweetest I've ever tasted.

When he died, my Swedish connection was severed. What was his homeland like, the landscapes, the people, the food? I knew I had to discover that part of my heritage for myself. Over the years, I had become a staunch environmentalist. A strong conservation ethic was part of Swedish culture, enabling a progressive, green country—all the more reason to visit.

Gazing at a postcard my mother had sent me thirty years before, when she visited a cousin of Grandpa's in Sweden, I decided to send Anders a letter, knowing full well that he had probably moved or died by that time. Was I in for a surprise.

Leaping out of bed one night at 3:45 A.M., I angrily grabbed the receiver and yelled "Hello!" That phone had been ringing since 3:00 A.M. Telemarketers at this hour? I had rolled over and tried to ignore the ringing until I couldn't stand it any longer.

"Sharon?" a man's voice asked from the other end. "This is Anders in Sweden." Struggling with his English on the phone, Anders said they wanted me to come visit them so they could show me where my grandfather was born. I was thrilled!

Six months later, I was sitting by the front door of a hotel in Växjö, a town near Tingsryd, where Anders lived. After about forty-five minutes, the door to a side hallway opened and a man dressed in tennis shoes, navy sweat pants, a tan cotton jacket, and a baseball cap entered. After some clumsy handshakes, Anders; his much-younger second wife, Susanna; their blonde four-year-old daughter, Dorotea; and I piled into his aging maroon Volvo.

Tingsryd is a town of 2,500, with well-kept neighborhoods, shops, and even a hotel. Anders pulled into the narrow driveway of a very small beige-colored house. Though Susanna has a yard-maintenance business, they own just one car. Once inside, I instantly began to scan the house for all things Swedish. In a corner of the cluttered living room, a traditional white ceramic-tile stove reached gracefully to the ceiling. A guitar against one wall was next to a white piano. Both were decorated with Swedish rosemaling, a style of folk-art painting. Anders was a high school music teacher until he retired a couple of years ago.

After coffee and biscuits (cookies), he pulled out a bulging album full of letters written in Swedish and old photos. Listening to him talk about the people in the photos, I got the feeling that he felt a tremendous sense of loss. "If we had more time, I could show you all the tools and things I have from our family in the garage."

After my grandfather died, I felt my Swedish roots died along with him. Three of Anders's children are grown and live in other parts of Sweden and his new wife is Hungarian; maybe he felt his own roots were a little thin. But there was no time to chat about the past since daylight was fading and he still wanted to take me to Tullanis, where Grandpa was born.

The Volvo chugged down narrow winding roads for about an hour before Anders turned at a small handwritten signpost nailed to a tree by the side of the road, driving down a dirt road into the pine forest. Soon we reached a clearing with about six traditional barn-red houses and one two-story house held together with concrete. He pulled up beside the two-story house and said it was the site where my grandfather lived before he came to America. The original house had burned down. The two-story house had nearly broken apart when it was moved to the new location a few years before my visit.

Sam, a young man who lived in the house with his family, came out to talk with me. "Your grandfather's father, mother, and later, his step-mother, had eleven children," he explained as we walked over the uneven ground around the house. "As you can see, this ground is full of rocks," he continued. "The soil was poor and it was hard to grow food. They were starving." Six brothers from the family eventually came to America.

Lumber piled in the yard attested to Sam's next project. "I'll send you a photo when I get the exterior done and painted red," he commented.

I might have been right at home growing up in Sweden, where my "protect the environment" views are woven into the fabric of their culture. The first country to establish an Environmental Protection Agency, in 1967, today Sweden is considered the greenest country on the planet. More than half of Sweden's energy supply comes from renewables.

According to Johan Rockström, author of *Big World, Small Planet*, in 1990 Sweden imposed a carbon tax of US$100 per ton on carbon dioxide emissions. Robert Charlson, University of Washington professor emeritus of atmospheric science, commented at a talk I attended on human centered climate change that "Sweden gets by with one-half the energy per capita as does the U.S." His suggestion to help the United States catch up? A tax at the gas pump.

Two branches of the same tree, with roots running deep. My "tight genes" helped me connect to my grandfather's birthplace. Caring for the environment and the Swedish "just enough, save some for others" ethic are ingrained in the culture. Somehow, I think America missed the boat.

WATERY WORLD

Weightless, suspended in silence, my breath was the only sound as I floated in the bathtub-warm turquoise water at Trunk Bay on St. John in the U.S. Virgin Islands. Through the snorkel mask, I watched curious angelfish, parrotfish, butterfly fish, and others dart around the coral, their brilliant blue, green, orange, yellow, red, and purple skins shimmering in the sunlit waters. Intrigued by the undersea wonderland around me, I nearly forgot my fear of water.

As long as I can remember, I've always been afraid of water. Swimming classes in chlorine-smelling indoor pools only made my fear worse. A skinny kid with no upper body strength, I'd kick ferociously, arms flailing, as I was sinking toward the bottom of the pool. Panic!

But this warm, turquoise Caribbean water was different—a peaceful place where exotic fish curiously darted around me. As long as I didn't go deeper than about four feet, I enjoyed it.

The U.S. Virgin Islands of St. Thomas, St. Croix, and St. John have lured visitors for centuries. Arawak and Carib Indians, pirates, and European explorers, including Christopher Columbus, all have left their marks. Remnants of sugar mills, part of the legacy of Dutch entrepreneurs in the eighteenth century, are still scattered around the islands.

Purchased from Denmark in 1917, the Virgin Islands remained relatively undisturbed until the 1950s, when development and tourism in the Caribbean rapidly increased. Tourism is still rising with more than a million cruise-ship passengers visiting St. Thomas every year. Some of those tourists take the small ferry over to St. John.

Virgin Islands National Park, established in 1956, covers just over half, or 12,900 acres, of the island and 5,600 acres of offshore waters. Despite commercial development on neighboring islands, St. John remains a sleepy green island ringed with white sandy beaches. Fragrances from red flamboyant blossoms along with pink and white frangipani perfume the air.

Sometimes two cruise ships a day, each carrying 2,500 to 6,500 passengers, dock at St. John, especially during the winter cruising season. The more I explored this lush island, the more I began to wonder what impact these hordes of tourists were having on the fragile environment.

Temperatures hover around eighty humid degrees year-round, with occasional brief showers September through January. A tropical breeze usually keeps the temperature tolerable in this perfect tropical island vacation spot. But how many visitors are aware that they need to tread lightly and take care not to damage the resources or they will destroy what they came to see?

Many of those sun-seeking tourists rent snorkel gear, then wade out into the warm ocean to explore the undersea world. Heeding advice on printed material at the visitor center, some are very careful not to step on the coral. Others, caught in the excitement of the moment, step on any solid foothold they find that will propel them farther along in their underwater exploration.

"Coral is easily broken and very slow growing," explained Chuck Weikert, former chief of interpretation for the park. "Snorkelers should avoid touching, standing on, or kicking coral with their fins," he continued. "Not only will this protect the coral reef, but it will also protect snorkelers from nasty cuts and scrapes from the coral."

Coral can become stressed in unusually warm water. Rising global temperatures warm the waters, causing coral bleaching and creating an environment where aquatic diseases can flourish. When the water gets too warm, coral expels the algae that gives it color. In 2005, 2010, and again in 2012, there were spikes in water temperature causing bleaching disease that left stands of grayish-whitish coral devoid of life.

The coral reefs face other hazards as well. According to a study in the Archives of Environmental Contamination and Toxicology, oxybenzone, a chemical found in more than 3,500 sunscreens worldwide may contribute to coral bleaching. The scary part is that all it takes is a drop of sunscreen in an area of water equivalent to the size of six Olympic swimming pools.

Coral reefs consisting of stony corals, sponges, and algae provide services unlike any other ecosystem. The reefs support a diversity of marine life including sea turtles, conchs, and lobsters. Juvenile fish find hiding places in the nooks and crannies of the coral. Even in this highly technological age when nearly anything can be created artificially, pharmaceutical researchers still comb the various elements of the reef for the next new drug. Coral reefs, along with mangrove stands, also provide shoreline protection in the event of tropical storms.

Coral reef habitats are valuable to tourism, but much more. The National Oceanic and Atmospheric Administration reports the value worldwide to fisheries is $100 million. Coral reefs provide a net benefit of $9.6 billion each year from tourism and recreation revenues, and $5.7 billion per year from fisheries.

Scientists in cooperation with the National Park Service are monitoring coral reefs in five South Florida and Caribbean Island parks.

Ranger-led hikes on St. John pass through lush tropical forests harboring eight hundred species of plants. Over 250 aging structures tell the story of the sugar-plantation era. Danish brick, brain coral, and other natural building materials are still clearly visible. Rangers remind visitors to stay on the trails and be careful around the historical structures, some of which are in need of repair.

One of my favorite stops was at Annaberg Sugar Plantation to watch the demonstrations of traditional island skills such as basket weaving, cooking with charcoal, and subsistence gardening. Annaberg was one of the twenty-five mills producing molasses and rum on St. John beginning in 1717. Slaves planted, cut, and processed the sugar cane. Emancipation of the Danish West Indies slaves in July 1848 and development of the beet-sugar extraction process heralded the end of the sugar-cane era in the Caribbean.

Educating visitors to respect the resources is one of the main concerns of National Park Service staff. Information about the resources is available at the visitor center, but some people just glance at the displays. Finding out about the best beaches is often their top priority.

Cruise ships bring thousands of visitors each year. Perhaps if education about the fragile ecosystems, both land and sea, that they are about to enter were stressed right on the cruise ship, people might pay more attention.

Tropical treasures such as those on St. John are too special to lose.

WINDS OF CHANGE

"How long have you had these blisters on your toes?" the doctor in the medical clinic on St. John asked me. Clad in a white lab coat, he inspected my toes carefully. A wrinkled brow told me he was concerned about those blisters.

"Oh, about a week," I responded.

"And you are just coming in to the clinic now?" he inquired. "You should have come in right away. These look like fire ant bites. You could have had a serious allergic reaction from these nasty bites."

My husband and I enjoyed our visit to St. John two years earlier, so we decided to return, taking time on this trip to explore Tortola, in the British Virgin Islands, before coming to St. John National Park in the U.S. Virgin Islands. Sitting in a lawn chair on the ground-floor patio of our hotel in Tortola, I felt tiny stings when the ants bit me, but didn't think much about it. Later that day they started to itch and blisters appeared.

When the ugly yellowish blisters weren't any better after a week, I went to the medical clinic in Cruz Bay a day or so after we arrived in St. John. Jim and I were there as volunteer photographers for the head of interpretation at the national park. Jim was taking color photos, and I was taking black-and-white images. But before we explored the island, I had to get my toes checked out.

The doctor told me to take an antihistamine once a day and soak my feet in warm water with Epsom salts. If they weren't gone in a week, he wanted me to come back to the clinic.

According to the Department of Agriculture, red fire ants (*Solenopsis richteri* and *Solenopsis invicta*) are tiny, aggressive insects with a big bite. Native to South America, they now inhabit the southern states, including the southern portions of Arizona, Wyoming, New Mexico, and California. As our environment warms, plants, animals, birds, and insects, including fire ants, are moving northward. The thought that the fire ants are working their way north to my home state of Washington is creepy.

Coming from our cool maritime climate in Seattle to tropical St. John in July, we had to quickly adjust to hot, humid weather. Island breezes, when they came, were welcome. After getting used to moving at a slower pace with soaking-wet shirts sticking to our backs, we adapted and got into a routine of packing up photography gear and lunch each morning for our photo adventures. Soon we were photographing white sandy beaches, tropical forests, historic remnants of the sugar-cane era, and tourists enjoying it all.

A tropical paradise with picture-perfect beaches and warm water, plus a rich history—a perfect vacation spot. At least it was when we visited in 1994. Yet all that changed on September 6, 2017, when Hurricane Irma smashed into the island with 175-mile-per-hour winds causing billions of dollars in damage. About three weeks later, Hurricane Maria dealt another blow to the island. Both were Category 5 hurricanes.

Luckily, there was enough advance warning before Hurricane Irma hit for many of the island residents to evacuate to neighboring islands which were not in the path of the hurricane.

Trees were snapped in half and blown down, and roads closed. Forty-five leisure craft were blown up on national park beaches or sunk in national park waters. Homes and businesses, including hotels, were damaged, some beyond repair. A few of the historic structures, such as the archaeology museum at Cinnamon Bay, were totally destroyed, while others were untouched.

Shortly after Hurricane Irma hit, crews from other national parks, the Fish and Wildlife Service, along with other agencies from the U.S. mainland were on the ground clearing roads, beaches, and demolished structures around the island. With the national park supporting the island economy, cleaning up and reopening the beaches and a few hotels in time for winter visitors was a top priority. By December 7, 2017, cruise ships were landing at St. Thomas and St. Croix, with ferries shuttling some of the tourists to a partially restored St. John.

St. John rarely lies in the path of hurricanes; the last one was in 1999. However, climate change is beginning to affect ecosystems on land and fuel more violent weather events. As storms pass over a warmer ocean, they pick up more energy, making winds much stronger and capable of catastrophic damage.

It will take years for the island to fully recover. I'm glad I got a chance to experience the natural beauty of St. John before the hurricanes did so much damage. I hope we are inspired to take steps now to reduce rising greenhouse gases that are changing or destroying fragile ecosystems in these special places.

RESOURCES

Below is a list of organizations whose top priorities include strategies to address climate change and protect the environment.

ENVIRONMENT

350 ~ *350.org*
Cascadia Climate Action ~ *cascadiaclimateaction.org*
Climate Reality Project ~ *climaterealityproject.org*
Climate Solutions ~ *climatesolutions.org*
Environmental Defense Fund ~ *edf.org*
Greenpeace ~ *greenpeace.org*
National Audubon Society ~ *audubon.org*
National Resources Defense Council ~ *nrdc.org*
Sierra Club ~ *sierraclub.org*
The Nature Conservancy ~ *nature.org*
World Watch Institute ~ *worldwatch.org*

EXTREME WEATHER

Climate Communication ~ *climatecommunication.org*
National Climate Assessment ~ *nca2014.globalchange.gov*
Union of Concerned Scientists ~ *ucsusa.org*

FOOD SUPPLY

Food and Agriculture Organization of the United Nations ~ *fao.org*
World Food Program USA ~ *wfpusa.org*

HUMAN HEALTH

U.S. Global Change Research Program ~ *globalchange.gov*
World Health Organization ~ *who.int*

MARINE MAMMALS

American Cetacean Society ~ *acsonline.org*

The Marine Mammal Center ~ *marinemammalcenter.org*

OCEANS

Ocean Conservancy ~ *oceanconservancy.org*

The Ocean Foundation ~ *oceanfdn.org*

OVERPOPULATION

Conserve Energy Future ~ *conserve-energy-future.com*

Population Connection ~ *populationconnection.org*

ABOUT THE AUTHOR

Sharon Sneddon has been a freelance nonfiction writer for over twenty-five years. Many of her articles about travel and natural history have appeared in local and national publications, including in-flight magazines for Alaska and TACA airlines, *National Parks* and *Northwest Travel* magazines, as well as the *Seattle Post-Intelligencer* and *Boston Herald* newspapers. For several years, she served as newsletter editor for the Seattle Audubon Society.

CPSIA information can be obtained
at www.ICGtesting.com
Printed in the USA
BVHW07s1156071018
529438BV00002B/6/P